High Performance
Liquid Chromatography

10/4/06

Titles in Series:

High Performance Liquid Chromatography

Analytical Chemistry by Open Learning

Author:
SANDY LINDSAY
East Ham College

Editor:
DAVID KEALEY

on behalf of ACOL

Published on behalf of ACOL, Thames Polytechnic,
London
by
JOHN WILEY & SONS
Chichester · New York · Brisbane · Toronto · Singapore

© Crown Copyright, 1987

Published by permission of the Controller of
Her Majesty's Stationery Office

Library of Congress Cataloging in Publication Data:

Lindsay, Sandy
    High performance liquid chromatography.
    (Analytical chemistry by open learning)
    1. High performance liquid chromatography—
    Programmed instruction.   2. Chemistry, Analytic—
    Programmed instruction.   I. Kealey, D. (David)
    II. ACOL (Firm : London, England) III. Title.
    IV. Series.
    QD79.C454L54   1987   543'.0894'077   87-8158

    ISBN 0 471 91372
    ISBN 0 471 91373 (pbk.)

British Library Cataloguing in Publication Data:

Lindsay, Sandy
    High performance liquid chromatography.
    — (Analytical chemistry)
    1. Liquid chromatography
    I. Title   II. Kealey, D.   III. ACOL   IV. Series
    543'.0894   QD79.C454

    ISBN 0 471 91372 3
    ISBN 0 471 91373 1 (pbk.)

Printed and bound in Great Britain

Analytical Chemistry

This series of texts is a result of an initiative by the Committee of Heads of Polytechnic Chemistry Departments in the United Kingdom. A project team based at Thames Polytechnic using funds available from the Manpower Services Commission 'Open Tech' Project has organised and managed the development of the material suitable for use by 'Distance Learners'. The contents of the various units have been identified, planned and written almost exclusively by groups of polytechnic staff, who are both expert in the subject area and are currently teaching in analytical chemistry.

The texts are for those interested in the basics of analytical chemistry and instrumental techniques who wish to study in a more flexible way than traditional institute attendance or to augment such attendance. A series of these units may be used by those undertaking courses leading to BTEC (levels IV and V), Royal Society of Chemistry (Certificates of Applied Chemistry) or other qualifications. The level is thus that of Senior Technician.

It is emphasised however that whilst the theoretical aspects of analytical chemistry can be studied in this way there is no substitute for the laboratory to learn the associated practical skills. In the U.K. there are nominated Polytechnics, Colleges and other Institutions who offer tutorial and practical support to achieve the practical objectives identified within each text. It is expected that many institutions worldwide will also provide such support.

The project will continue at Thames Polytechnic to support these 'Open Learning Texts', to continually refresh and update the material and to extend its coverage.

Further information about nominated support centres, the material or open learning techniques may be obtained from the project office at Thames Polytechnic, ACOL, Wellington St., Woolwich, London, SE18 6PF.

How to Use an Open Learning Text

Open learning texts are designed as a convenient and flexible way of studying for people who, for a variety of reasons cannot use conventional education courses. You will learn from this text the principles of one subject in Analytical Chemistry, but only by putting this knowledge into practice, under professional supervision, will you gain a full understanding of the analytical techniques described.

To achieve the full benefit from an open learning text you need to plan your place and time of study.

- Find the most suitable place to study where you can work without disturbance.

- If you have a tutor supervising your study discuss with him, or her, the date by which you should have completed this text.

- Some people study perfectly well in irregular bursts, however most students find that setting aside a certain number of hours each day is the most satisfactory method. It is for you to decide which pattern of study suits you best.

- If you decide to study for several hours at once, take short breaks of five or ten minutes every half hour or so. You will find that this method maintains a higher overall level of concentration.

Before you begin a detailed reading of the text, familiarise yourself with the general layout of the material. Have a look at the course contents list at the front of the book and flip through the pages to get a general impression of the way the subject is dealt with. You will find that there is space on the pages to make comments alongside the text as you study—your own notes for highlighting points that you feel are particularly important. Indicate in the margin the points you would like to discuss further with a tutor or fellow student. When you come to revise, these personal study notes will be very useful.

∏   When you find a paragraph in the text marked with a symbol such as is shown here, this is where you get involved. At this point you are directed to do things: draw graphs, answer questions, perform calculations, etc. Do make an attempt at these activities. If necessary cover the succeeding response with a piece of paper until you are ready to read on. This is an opportunity for you to learn by participating in the subject and although the text continues by discussing your response, there is no better way to learn than by working things out for yourself.

We have introduced self assessment questions (SAQ) at appropriate places in the text. These SAQs provide for you a way of finding out if you understand what you have just been studying. There is space on the page for your answer and for any comments you want to add after reading the author's response. You will find the author's response to each SAQ at the end of the text. Compare what you have written with the response provided and read the discussion and advice.

At intervals in the text you will find a Summary and List of Objectives. The Summary will emphasise the important points covered by the material you have just read and the Objectives will give you a checklist of tasks you should then be able to achieve.

You can revise the Unit, perhaps for a formal examination, by re-reading the Summary and the Objectives, and by working through some of the SAQs. This should quickly alert you to areas of the text that need further study.

At the end of the book you will find for reference lists of commonly used scientific symbols and values, units of measurement and also a periodic table.

Contents

Study Guide

High performance liquid chromatography (hplc) is the most powerful of all the chromatographic techniques. It can often easily achieve separations and analyses that would be difficult or impossible by other forms of chromatography. On the other hand, there are very many things that can go wrong with the separation; there are probably more pitfalls in hplc than in any other form of chromatography. To avoid these, you have to have the sort of experience that is difficult to obtain by reading textbooks. Only by doing a great deal of experimental work (and making many mistakes) can you hope to achieve the necessary practical skills.

This is not to say that the theoretical side of the subject is unimportant. In chromatography, theory has always led experimental work. The great advances that have been made in liquid chromatography in the last 10–15 years have been achieved through a better theoretical understanding of the technique. You will not be able to use hplc to full advantage unless you have a proper understanding of how it works. Liquid chromatography is a very wide ranging subject and to understand it you will have to have some knowledge of many different areas of physical and analytical chemistry. I will assume you have studied chemistry up to the standard of the BTEC Higher Certificate, and that you have a knowledge of physics and mathematics to about 'O' level standard. I will also assume that you are familiar with *ACOL: Chromatographic Separations*. You will find it helpful to have had some experience with the use of analytical instruments such as spectrophotometers and chart recorders.

Because hplc covers such a wide area, you are bound to find that there are some topics that you would like to study in more detail than is given in this text. I would recommend the books by Knox or by Hamilton and Sewell as a starting point (details of these are given in the bibliography). Because hplc is a technique that is still developing, textbooks often contain some material that is obsolete. To get up to date information, especially on instrumentation, you have to use the chromatographic literature, or catalogues and applications literature from equipment manufacturers. Several manufacturers supply quite informative catalogues; the Chrompack *Guide to Chromatography* and the Waters *Sourcebook* are good examples.

Supporting Practical Work

1. GENERAL CONSIDERATIONS

The experiments below use reverse phase chromatography with bonded silica columns and uv absorbance detection. If more extensive experimental facilities are available, some additional experiments are suggested. These are concerned with the preparation and evaluation of columns, and with the use of other detectors and modes of hplc. It should be possible to complete each experiment within a three hour practical period.

2. AIMS

(*a*) To provide practical experience in the use of basic hplc equipment.

(*b*) To demonstrate the various parameters that control hplc separations.

(*c*) To show the use of the technique for separation and quantitative analysis.

(*d*) To illustrate some of the important principles from the theory part of the Unit.

3. SUGGESTED EXPERIMENTS

(*a*) The effect of mobile phase flow rate and dead volume on column performance.

(*b*) The effect of mobile phase composition on retention and selectivity in a reverse phase separation.

(*c*) Determination of 4-hydroxy-3-methoxy benzaldehyde (vanillin) in vanilla essence.

(*d*) Determination of aspirin and caffeine in an analgesic tablet.

4. ADDITIONAL EXPERIMENTS

(*a*) Preparation and evaluation of an hplc column.

(*b*) Analysis of sugars in fruit juice.

(*c*) The use of extraction techniques in the separation of carotene pigments from fruit.

Bibliography

1. Textbooks on high performance liquid chromatography

(*a*)   J J Kirkland and L R Snyder, *Introduction to Modern Liquid Chromatography*, 2nd Edn, John Wiley, 1979.

(*b*)   R J Hamilton and P A Sewell, *Introduction to High Performance Liquid Chromatography*, Chapman and Hall, 1982.

(*c*)   *High Performance Liquid Chromatography*, Ed. J H Knox, Edinburgh University Press, 1982.

Reference (*a*) is a very comprehensive treatment.
References (*b*) and (*c*) are more suited to this text.

2. Additional reading for specialised topics

Small bore columns

(*d*)   *Small bore columns in Liquid Chromatography*, Ed. R P W Scott, John Wiley, 1984.

(*e*)   *Techniques in Liquid Chromatography*, Ed. C F Simpson, John Wiley, 1984. Chapter 4.

Post column reactors

(*f*)   R W Frei, H Jansen, U A T Brinkman, *Analytical Chemistry*, 1985, 57(14) 1529A-1539A.

Separation of chiral compounds

(*g*)   R Audebert, *J Liquid Chromatography*, 1979, 2(8), 1063–1095.

(*h*)   D Johns, *International Laboratory* November/December 1985, 32–39.

Column switching

(*j*)   C J Little, O Stahel, W Lindner, R W Frei, *American Laboratory* (1984), 16(10), 120–2, 125–9.

(*k*)   R E Majors, *J. Chromatographic Science* (1980), **18**, 571–579.

Practical methods with columns and samples

(*l*)   F M Rabel, *J. Chromatographic Science* (1980), **18**, 394–408.

(*m*)   Reference (*e*), Chapter 3.

Acknowledgements

Figure 2.2f is based on *Introduction to HPLC*, R. J. Hamilton and P. A Sewell, Chapman and Hall, 1982, Figure 3.10.

Figure 2.2g is redrawn from *Chrompack 1981* catalogue with the permission of Chrompack Ltd.

Figure 2.3i is redrawn from *Supelco HPLC bulletin 819*. Permission applied for.

Figure 2.4e is redrawn from *Philips Analytical LC Systems 1986*, with the permission of Pye Unicam Ltd.

Figures 2.4h, 2.4o, 3.2f and 3.3g are redrawn from Perkin Elmer Publications. Permission has been requested.

Figures 3.1a, 4.3m and 4.3n and redrawn from *Practical High Performance Liquid Chromatography*, Ed. C. Simpson, Heyden, 1976, with permission of John Wiley & Sons Ltd.

Figure 3.2d is redrawn from *Waters Source Book 1986*, with the permission of the Millipore Corporation.

Figure 3.2e is redrawn from *Journal of Chromatographic Science*, 18, 519, 1980. Permission has been requested.

Figures 3.3d, 3.3e, 3.3f, 3.4a, 3.4b, 3.4g, 4.2b-4.2f, 4.3c, 4.3f-4.3j are redrawn from *Waters LC School* with the permission of the Millipore Corporation.

Figure 3.1g is redrawn from *Journal of Chromatography* 158, 233, 1978. Permission has been requested.

1. Introduction

High performance liquid chromatography is a technique that has arisen from the application to liquid chromatography (lc) of theories and instrumentation that were originally developed for gas chromatography (gc).

Classical liquid chromatography has been around for quite a long time, and you will probably have used it in one form or another. In the original method an adsorbent, for instance alumina or silica, is packed into a column and is eluted with a suitable liquid. A mixture to be separated is introduced at the top of the column and is washed through the column by the eluting liquid. If a component of the mixture (a solute) is adsorbed weakly onto the surface of the solid stationary phase it will travel down the column faster than another solute that is more strongly adsorbed. Thus separation of the solutes is possible if there are differences in their adsorption by the solid. This method is called *adsorption chromatography or liquid solid chromatography* (lsc).

In lc there are other sorption mechanisms that can cause separation, depending on whether we choose to use a liquid or a solid as the stationary phase, or what kind of solid we use. *Liquid–liquid chromatography* (llc) uses a liquid stationary phase coated onto a finely divided inert solid support. Separation here is due to differences in the partition coefficients of solutes between the stationary liquid and the liquid mobile phase. In normal phase llc the stationary phase is relatively polar and the mobile phase relatively non-polar, whilst reverse phase llc uses a non-polar stationary liquid and a polar mobile phase.

In *ion-exchange chromatography*, the stationary phase is an ion-exchange material, usually a resin, and separations are governed by the strength of the interactions between solute ions and the exchange sites on the resin. Finally, in *exclusion chromatography*, the stationary phase is a wide pore gel that can separate molecules on the basis of their size and shape, the largest molecules travelling most rapidly through the system.

Experimentally, classical lc was done by packing the stationary phase into a glass column, maybe 5 cm in diameter and 1 m in length, and eluting with a suitable solvent, or range of solvents. The column could often be used only once, having to be repacked for each sample that was examined. In llc the eluting solvent had to be saturated with the stationary liquid phase in order to avoid stripping the stationary liquid from the column. Many of the stationary phases used were not very efficient, so that for tricky separations long columns had to be used, the separations took a long time and used large amounts of solvent. The separated solutes were isolated by dividing the output of the column up into a series of arbitrary fractions which were then evaporated down so that any solutes present could be identified by other physical or chemical means (eg melting point, elemental analysis, spectrometry).

The development of the open-column methods, ie paper chromatography (in the 1940's) and thin-layer chromatography (in the 1950's) greatly improved the speed and resolution of lc, but there were still serious limitations compared to modern lc methods, in that analysis times were long, resolution was poor and quantitative analysis, preparative separations and automation were difficult.

It was known from gas chromatographic theory that efficiency could be improved if the particle size of the stationary phase materials used in lc could be reduced. High performance liquid chromatography developed steadily during the late 1960s as these high efficiency materials were produced, and as improvements in instrumentation allowed the full potential of these materials to be realised. As hplc has developed, the particle size of the stationary phase used has become progressively smaller. The stationary phases used today are called *microparticulate* column packings and are commonly uniform, porous silica particles, with spherical or irregular shape, and nominal diameters of 10, 5 or 3 μm. The different separation mechanisms mentioned earlier can be realised by bonding different chemical groups to the surface of the silica particle to produce what are called *bonded phases*. Chromatography suppliers list a variety of these, but about 75% of the work in hplc at the moment is done using a bonded phase in which C-18 alkyl groups are attached to the surface of the silica particles. These types are called ODS (octadecylsilane) bonded phases. With bonded phases, the nature of the sorption mechanism is sometimes not clear, and there is much theoretical and experimental work going on at the moment attempting to clarify such mechanisms.

When packed into a column, the small size of these particles leads to a considerable resistance to solvent flow, so that the mobile phase has to be pumped through the column under high pressure. Typically, the column is 10–25 cm long and 4.6 mm internal diameter. Although these columns are expensive, they are re-usable, so that the cost can be spread over a large number of samples. The column and all the associated plumbing must be able to withstand the pressures that are used and must also be chemically resistant to the mobile phase solvents. Columns are usually made of stainless steel, although glass or plastics are favoured by some manufacturers. At the moment there is considerable interest in the properties of columns that have a diameter of 2 mm or less (known as small bore or microbore columns), and it is possible that these may become widely used in the future.

In analytical hplc, the mobile phase is pumped through the column at a flow rate of 1–5 cm^3 min^{-1}. If the composition of the mobile phase is constant, the method is called *isocratic* elution. Alternatively, the composition of the mobile phase can be made to change in a predetermined way during the separation, which is a technique called *gradient* elution. Gradient elution is used in situations similar to those requiring temperature programming in gc, and is necessary when the range of retention times of solutes on the column is so large that they cannot be resolved and eluted in a reasonable time using a single solvent or solvent mixture. In adsorption chromatography, for instance, non-polar solutes are adsorbed relatively weakly and should be eluted with a non-polar solvent, whereas polar solutes are adsorbed more strongly and require a more polar solvent. If the sample contains a wide range of polarities, the separation could be done by changing the polarity of the solvent mixture during the separation. In other cases it may be necessary to use gradient elution where other properties of the solvent (eg pH or ionic strength) are changed.

After passing through the column, the separated solutes are sensed by an in-line detector. The output of the detector is an electrical signal, the variation of which is displayed on a potentiometric recorder, a computing integrator or a vdu screen. Most of the popular detectors in hplc are selective devices, which means that they may not respond to all of the solutes that are present in a mixture. At present there is no universal detector for hplc that can compare with the sensitivity and performance of the flame ionisation detector used in gas chromatography. Some solutes are not easy to detect in hplc, and have to be converted into a detectable form after they emerge from the column. This approach is called *post-column derivatisation*.

As in other forms of chromatography, the time taken for the solute to pass through the chromatographic system (the retention time) is a characteristic of the solute for a particular set of conditions. However, to use retention data on its own for the identification of unknown solutes would be rather like trying to identify an unknown organic compound simply by measuring its melting or boiling point. Many different solutes will have identical retention times for a particular set of conditions. Chromatography is an excellent method for the separation of mixtures, but it does not provide the detail necessary for the clear identification of the separated compounds. Such detail is provided by spectrometric techniques so it is not surprising that much effort has been expended to try and combine them with hplc. For example, some detectors can record and store the uv spectra of solutes as they emerge from the column. A much more powerful (and very expensive) method is the direct combination of liquid chromatography and mass spectrometry. In both cases, modern data processing methods allow you to match the spectra obtained with the spectra of standard substances, obtained from libraries of recorded spectra.

How do we decide whether to separate a mixture by gc or hplc? In gc, mixtures are examined in the vapour phase, so that we have to be able to form a stable vapour from our mixture, or convert the substances in it to derivatives that are thermally stable. Only about 20% of chemical compounds are suitable for gc without some form of sample modification; the remainder are thermally unstable or involatile. In addition, substances with highly polar or ionisable functional groups often show poor chromatographic behaviour by gc, being very prone to tailing. Thus hplc is the better technique for macromolecules, inorganic or other ionic species, labile natural products, pharmaceutical compounds and biochemicals.

In gc there is only one phase (the stationary liquid or solid phase) that is available for interaction with the sample molecules. Because the mobile phase is a gas, all sample vapours are soluble in it in all proportions. In hplc both the stationary phase and the mobile phase can interact selectively with the sample. Interactions such as complexation or hydrogen bonding that are absent in the gc mobile phase may occur in the hplc mobile phase. The variety of these selective interactions can also be increased by suitable chemical modification of the silica surface, hence hplc is a more versatile technique than gc, and can often achieve more difficult separations.

There will often be areas where either technique could be used, and in such cases gc is usually chosen. One reason for this is that hplc tends to be a more expensive technique than gc, both in capital outlay for the instrument and in day-to-day running costs. The gc separation would also probably be faster and more sensitive.

SAQ 1a

Complete the following definition of liquid chromatography by filling in the blanks. For each space, choose a word from the groups given below:

Liquid chromatography is a technique for the .... of mixtures in which the sample is introduced into a system of two .... Differences in .... shown by the solutes cause them to travel at different speeds in the ....

(*i*)	analysis separation determination	(*iii*)	adsorption distribution partition
(*ii*)	substances chemicals phases	(*iv*)	liquid mobile phase system

SAQ 1b

For which of the following would hplc be a suitable means of analysis?

(*i*) Determination of the composition of cigarette lighter fuel.

(*ii*) Analysis of ascorbic acid (vitamin C) in a vitamin C tablet.

(*iii*) Determination of the amount of caffeine in a soft drink.

(*iv*) Separation of a mixture of naturally occurring sugars.

(*v*) Separation of a mixture of amines.

Summary

A brief description is given of the way in which modern liquid chromatography has been developed from classical techniques. The important components of a high performance liquid chromatograph are introduced and the method is compared with gas chromatography as a separation technique.

Objectives

You should now be able to:

• list the mechanisms by which separations are achieved in hplc;

• identify the basic components of a high performance liquid chromatograph;

• appreciate that a very wide range of samples can be separated using the technique.

2. Instrumentation

2.1. THE BASIC COMPONENTS OF A HIGH PERFORMANCE LIQUID CHROMATOGRAPH

You will have seen from reading the introduction that an hplc instrument requires a high pressure pump and a supply of mobile phase, a column packed with a high efficiency stationary phase, an injection unit for introducing our samples on to the column, an in-line detector and some method of displaying the detector signal. Fig. 2.1 is a block diagram showing the way in which these different components are arranged to form a high performance liquid chromatograph.

Any part of the system that is in contact with the mobile phase must be made of materials that are not attacked by any of the solvents that are to be used. The wetted parts are usually made of stainless steel or ptfe although other materials, such as sapphire, ruby, or ceramics are sometimes used. Everything on the high pressure side, ie from the pump outlet to the end of the column, must be strong enough to withstand the pressures involved.

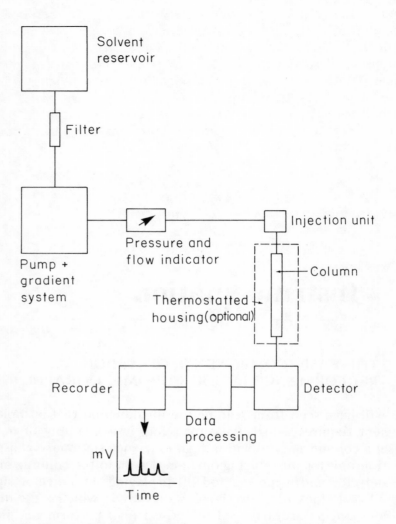

Fig. 2.1. *Block diagram of a high performance liquid chromatograph*

Another very important design consideration is that between the point at which the sample is introduced and the point at which it is detected, the dead volume in the equipment must be kept to a minimum. Dead volume means any empty space or unoccupied volume. The presence of too much dead volume can lead to disastrous losses in efficiency (Fig. 2.3c and 5.3b show examples of this). Clearly there will be some dead volume in the column itself, which will be the space that is not occupied by the particles of stationary phase.

∏   Can you list the other sources of dead volume?

     The other sources of dead volume are:

     (*a*) The injection unit

     (*b*) The tubing and fittings connecting the injection unit to the column

     (*c*) The tubing and fittings connecting the end of the column to the detector

     (*d*) The detector cell.

We should therefore use small bore tubing in short lengths for making the injector-column and the detector-column connections, and the injector and detector must be designed so that their internal volume is as small as possible. Dead volume before the introduction of the sample should also be minimised, to facilitate rapid changes of mobile phase composition that may be required during gradient elution.

We will now look at some of the different parts of the chromatograph in a little more detail.

2.1.1. The Mobile Phase Reservoir

In hplc the mobile phase can be an aqueous-organic mixture, a buffer solution or a mixture of organic solvents, depending on the chromatographic method and on the detector that is used. The simplest reservoir is a 1 dm^3 glass bottle, fitted with a lid and a 1/8 inch diameter ptfe tube to carry the mobile phase from the reservoir to the pump. It is important that the liquid entering the pump does not contain any dust or other particulate matter, as this can interfere with the pumping action and cause damage if it gets into the seals or valves. Such material can also collect on top of the column, causing irregular behaviour, or maybe even blockages. The mobile phase is therefore filtered before it enters the pump. This can be done by a stainless-steel filter element that is a push fit onto the end of the ptfe tube in the reservoir, or alternatively an in-line filter can be used. The filter size used is normally 2 μm.

It is also important to remove dissolved air or suspended air bubbles. Pockets of air can collect in the pump or the detector cell or in other places, causing strange behaviour from the detector and irregular pumping action (at worst, the pumping action can be lost completely). Practical methods for degassing the mobile phase are dealt with in Section 5.2. Some instruments have quite complicated arrangements for degassing the mobile phase in the reservoir; alternatively it can be degassed beforehand.

2.1.2. Pressure, Flow and Temperature

Column inlet pressures in hplc can be as much as 200 times atmospheric pressure, and hplc columns are packed using much larger pressures (up to 700 times atmospheric). The SI unit of pressure is the Pascal (1 Pa = 1 Nm^{-2}); normal atmospheric pressure is about 10^5 Pa. Because it is convenient to express pressure using reasonably small numbers, experimental workers and instrument manufacturers report pressures in bar, or pounds per square inch (psi), or sometimes in kg cm^{-2}. The bar is defined by 1 bar = 10^5 Pa, so that 1 bar corresponds roughly to normal atmospheric pressure. You will need to be able to convert between these units.

∏   See if you can work out the conversion between bar and psi, given that 1 pound = 0.4536 kg, 1 inch = 2.54 cm, and g = 9.81 ms^{-2}.

1 psi = a force of 0.4536 × 9.81 N acting over an area of 0.0254^2 m^2.

1 psi = 0.4536 × 9.81/(0.0254)^2 = 6897 Pa

1 bar = 10^5/6897 = 14.5 psi

The conversions for kg cm^{-2} are:

1 kg cm^{-2} = 0.981 bar = 14.2 psi.

Although as mentioned before, the column inlet pressure can be as much as 200 bar, most of the work in analytical hplc is done using pressures between about 25 and 100 bar. The pressure developed will depend on the length of the column, the particle size of the stationary phase, and the viscosity and flow rate of the mobile phase. Because liquids are not very compressible there is not much energy stored in them at high pressures, and the pressures used in hplc do not represent a hazard (precautions should be taken when packing columns, when the pressures used are much higher). The pressures above correspond to mobile phase flow rates of roughly 1–5 cm^3 min^{-1} through the column. For constant flow pumps (Section 2.2.3) the mobile phase flow rate can be set at the pump. For some of the cheaper pumps the flow setting is not very reliable and will not equal the flow through the column if there are leaks anywhere in the system. The flow can be measured at the column outlet by collecting the liquid for a known time and weighing it or measuring the volume.

Many commercial hplc instruments optionally provide a forced air oven which will control temperature with a stability of typically ±0.1 °C from ambient temperature to 100 °C. Because of the use of flammable solvents, safety considerations are important, so the ovens are usually provided with a facility for nitrogen purging and are designed to prevent the build up of solvent vapour in the event of a leak. If temperature control is used, it is important that the sample and the mobile phase are at the right temperature before being introduced to the column, so the mobile phase is passed through a preheating coil before it reaches the injection point.

Temperature control is important for the accurate measurement of retention data, and has to be used with refractometer detectors (Section 2.4.5). Increasing the temperature can increase the speed of the separation, especially in exclusion chromatography, and usually increases the efficiency of the column (though the gain in efficiency can be lost if the mobile phase is not properly equilibrated). Complicated separations can often be optimised by increasing the temperature, but this is done very much on a trial and error basis, and most work in hplc is still done without temperature control.

Summary

The arrangement of the basic components in a high performance liquid chromatograph is described and the sources of dead volume in the instrument are indicated. Typical working pressures and flow rates are given.

Objectives

You should now be able to:

● understand the arrangement of the main components of a high performance liquid chromatograph;

● identify typical materials used in hplc equipment;

● specify pressures and flow rates used in hplc;

● recognise the need for minimal dead volume in the equipment, and identify the sources of dead volume.

2.2. PUMPS AND INJECTION SYSTEMS

2.2.1. Pumps: General Considerations

The function of the pump in hplc is to pass mobile phase through the column at high pressure and at a controlled flow rate. One class of pump (constant pressure pump) does this by applying a constant pressure to the mobile phase; the flow rate through the column is determined by the flow resistance of the column and any other restrictions between the pump and the detector outlet. Another type (constant flow pump), generates a given flow of liquid, so that the pressure developed depends on the flow resistance.

The flow resistance of the system may change with time; this can be caused by swelling or settling of the column packing, small changes in temperature, or the build up of foreign particulate matter from samples, pump or injector. If a constant pressure pump is used, the flow rate will change if the flow resistance changes, but for constant flow pumps changes in flow resistance are compensated for by a change of pressure. Small flow changes are undesirable, as they will cause retention data to lack precision, and may cause an erratic baseline on the recorder. Because of this, it is advisable not to use constant pressure pumps in hplc instruments, although they are suitable for packing columns, where small changes in flow do not matter.

In addition to being able to pump the mobile phase at high pressure and constant flow, the pump should also have the following characteristics:

(*a*) The interior of the pump should not be corroded by any of the solvents that are to be used

(*b*) A range of flow rates should be available, and it should be easy to change flow rate

(*c*) The solvent flow should be non-pulsing

(*d*) It should be easy to change from one mobile phase to another (*e*) The pump should be easy to dismantle and repair.

These require some additional comment:

(*a*) All of the wetted parts of the pump should be made of inert materials of the sort described in Section 2.1. Even with these materials the pump may not like some solvent systems. For instance, high concentrations of chloride ion or citrates will slowly corrode stainless steel.

(*b*) The pressures needed to achieve the flow rates required in analytical hplc are not likely to exceed 150–200 bar. Most pumps are rated for much higher pressures than these (see Fig. 2.2a). At high pressure, the flow rate through the column may be less than that selected, due to the small compressibility of the mobile phase, or small leaks.

(*c*) Some types of constant flow pump produce a pulsing flow of mobile phase. If the detector used is flow sensitive this may produce baseline noise on the chromatogram, the recorder pen describing a sawtooth trace that tracks the motion of the pump piston. These flow variations can be reduced using a pulse damper, which is compressible dead volume placed between the pump and injector. Sometimes the dead volume of the column itself is enough to give satisfactory damping; if not, a small coil of tube can be used. Large dead volumes between the pump and injector should, however, be avoided (see below).

(*d*) The internal volume of the pump and of all the plumbing between pump and injector should be kept as small as possible, and the system should not have any off-line recesses, otherwise changing the composition of the mobile phase will take a long time, as the new one has to sweep out the old. An example of an off-line recess is a Bourdon gauge teed into the line as a simple and cheap method to measure pressure (most instruments use flow through devices). This creates a recess that is not swept by the liquid flow. If the new mobile phase is immiscible with the original then the changeover has to be done via a third solvent that is miscible with each.

Even if great care is taken with the pump, seals, rings and gaskets sometimes have to be replaced, so they should be easy to get at. Although chemically resistant, the seals and rings are relatively soft, and are prone to wear. If the mobile phase has not been properly filtered, small particles can get trapped between the ball and the valve seat in a check valve, causing the valve seat to wear. During the operation of the pump, it is possible for solution to creep between the piston and the seal. When the pump is idle, this solution may evaporate, and if it contains dissolved solids (as in buffer solutions) a solid deposit will be left on the piston, and the seal will be damaged when the pump is used again. Bacteria can sometimes grow in buffers; if such material is pumped it may block the column, or even cause blockages in the pump, if it is fitted with internal filters. Buffer solutions must therefore be carefully removed after use, by pumping water or a suitable solvent for several minutes.

Fig. 2.2a lists some of the properties of three different types of pump (the operating principle of each is explained in the next section).

Model	Phillips PU 415	ISCO LC 500	Stanstead A9512 LC
Type	constant flow twin reciprocating	constant flow syringe type	constant pressure pneumatic amplifier
Maximum output pressure, bar	400	250	500
Flow rate range, cm^3 min^{-1}	0.01–9.9	$1.3 \times 10^{-4} - 3.34$	up to 200
Capacity, cm^3	continuous pumping	375	continuous pumping

Fig. 2.2a. *Operating properties of three different types of pump*

2.2.2. Constant Pressure Pumps

The earliest form of constant pressure pump in hplc (the coil pump) used pressurised gas from a cylinder to drive mobile phase from a holding coil through the column. This type of pump was used in some of the older hplc instruments, but is now only of historical interest. If you want to know any more about it, there are details in most textbooks.

The operating principle of the pneumatic amplifier pump is shown in Fig. 2.2b.

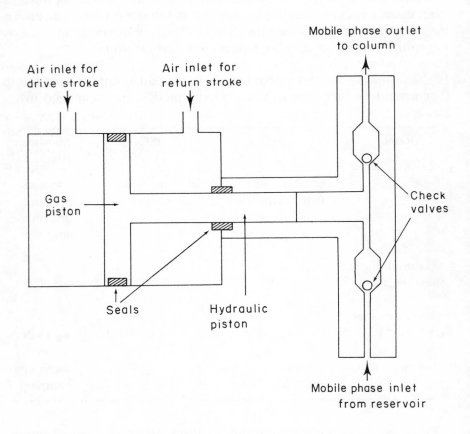

Fig. 2.2b.   *Pneumatic amplifier pump*

Air from a cylinder at pressures up to about 10 bar (150 psi) is applied to a gas piston that has a relatively large surface area. The gas piston is attached to a hydraulic piston that has a smaller surface area. The pressure applied to the liquid = gas pressure × area of gas piston/area of hydraulic piston. With 10 bar inlet pressure and a 50:1 area ratio, the hydraulic pressure obtained is 500 bar (7500 psi). On the drive stroke, the outlet valve on the pump head is open to the column and the inlet valve closed to the mobile phase reservoir. At the end of the drive stroke, the air in the chamber is vented and air enters on the other side of the gas piston to start the return stroke. On the return stroke the outlet valve closes, the inlet valve opens and the pump head refills with mobile phase. The pump can be started and stopped by operation of a valve fitted between the cylinder regulator and the pump.

Compared to syringe type or reciprocating pumps, pneumatic amplifier pumps are very cheap. They tend to be rather difficult to dismantle for repairs, and some types are very noisy in operation. Because they do not provide a constant flow of mobile phase, they are not used much in analytical hplc. They can, however, operate at high pressures and flow rates and so are used mainly for packing columns, where high pressures are needed and variations in the flow rate through the column do not matter.

2.2.3.   Constant Flow Pumps

Two types of constant flow pump have been used in hplc. Fig. 2.2c shows a syringe type pump.

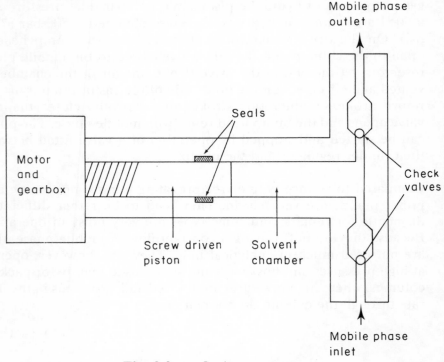

Fig. 2.2c.   *Syringe pump*

Mobile phase is displaced from a chamber by using a variable speed stepper motor to turn a screw which drives a piston. The chamber has a volume of 200–500 cm^3. The flow is pulseless, and can be varied by changing the motor speed. The mobile phase capacity is limited to the volume of the solvent chamber. Although this is fairly large, so that many chromatograms can be run before the chamber has to be refilled, a lot of solvent is wasted in flushing out the pump when a change is required.

The type of pump used in most instruments is the reciprocating pump, shown in Fig. 2.2d.

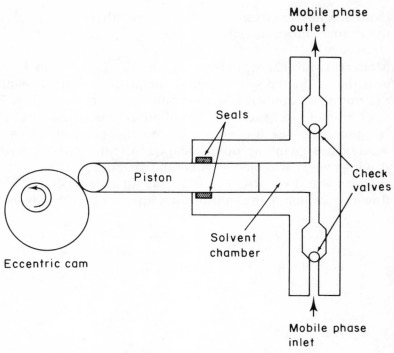

Fig. 2.2d. *Reciprocating pump*

The piston is driven in and out of a solvent chamber by an eccentric cam or gear. On the forward stroke, the inlet check valve closes, the outlet valve opens, and mobile phase is pumped to the column; on the return stroke the outlet valve closes and the chamber is refilled. Unlike syringe pumps, reciprocating pumps have an unlimited capacity, and their internal volume can be made very small, from 10–100 μl. The flow rate can be varied by changing the length of stroke of the piston or the speed of the motor. Access to the valves and seals is usually fairly straightforward.

In the single-headed reciprocating pump shown in the figure, the mobile phase is being delivered to the column for only half of the pumping cycle. During the drive stroke of the piston, the flow rate is not constant (because the speed of the piston changes with time). The output of the pump is shown in Fig. 2.2e (*i*). Use of a twin-headed pump with the two heads operated 180° out of phase (so that while one head is pumping the other is refilling), produces the output shown in (*ii*).

Modern twin-headed pumps use two pistons driven by a cam or gear that is shaped so as to make the piston speed constant. Ideally, the output of one such head should be as shown in Fig. 2.2e (*iii*), operation of two heads 180° out of phase producing a pulseless flow. In practice, on each drive stroke the change in flow rate is not instantaneous giving an output shown in (*iv*). To overcome this, the driving cam is arranged to make the piston travel faster on the refill than on the drive stroke, producing an output shown in (*v*). The flow rate, which is the sum of the output of both heads, is constant.

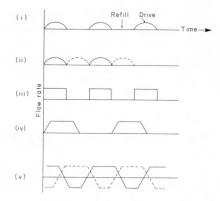

Fig. 2.2e. *Output from reciprocating pumps*

(*i*)   single headed pump
(*ii*) twin headed pump, heads 180° out of phase
(*iii*) single headed pump with constant speed (ideal)
(*iv*) single headed pump with constant piston speed (in practice)
(*v*)   twin headed pump, heads 180° out of phase, with different constant speeds on the drive and refill strokes.

Another way of reducing flow noise with single-headed pumps is to use a rapid stroke rate (one model uses 23 strokes s⁻¹) so that the detector cannot react rapidly enough to sense the flow changes. Many pumps also use feedback flow control, where the flow rate is measured downstream of the pump; any difference between the measured and the set flow actuates a change of motor speed so as to reduce the flow difference to zero.

2.2.4. Gradient Formers

Fig. 2.2f (*i*)–(*iii*) shows block diagrams of three types of gradient former. At low pressure, gradients can be formed from solvents A and B by metering controlled amounts of A into B into a mixing chamber before the high pressure pump (*i*). Alternatively, the composition in the low pressure mixing chamber can be controlled by using time proportioning valves as in (*ii*). This requires microprocessor control of the valves, which have to be switched very rapidly and accurately. In (*iii*), the separate solvents can be pumped with two high pressure pumps into a high pressure mixing chamber. The type of gradient formed is controlled by programming the delivery of each pump, but the use of two pumps is a very expensive means of forming solvent gradients.

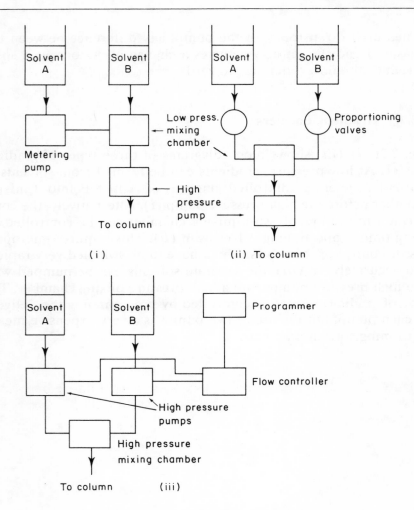

Fig. 2.2f.   *Gradient formers*

2.2.5.   Sample Injection

The earliest injection method in hplc used a technique borrowed
from gc in which a microlitre syringe was employed to inject the
sample through a self-sealing rubber septum held in an injection
unit at the top of the column. In another method, (stopped flow),
the flow of mobile phase through the column was halted and when

the column reached ambient pressure the top of the column was opened and the sample introduced at the top of the packing. These two methods are quite cheap, and are still used sometimes in home-made instruments. If you would like to read any more about them, there is a good account in the textbook edited by J. H. Knox.

Commercial chromatographs incorporate valves for injection. Although these are expensive, they are easy to use, give good precision, and are easily adapted for automatic injection. Sampling valves can be made with external or internal loops (sample holders). External loop valves have six ports, with a fixed volume sample loop connected across one pair. The other four ports are used to carry the mobile phase and the sample in and out of the valve. The loop is first filled with sample using a microlitre syringe, then on turning the valve from 'load' to 'inject' the contents of the sample loop are carried into the column by the mobile phase. External loop valves are used for relatively large samples, the smallest external loops being about 10 μl.

For smaller samples, four port valves with an internal loop can be used. In these, the sample loop is an engraved slot in the body of the valve. With both types, the volume of sample that has to be used to flush out and fill the loop is about ten times the loop volume. Fig. 2.2g shows the operating principle of each type.

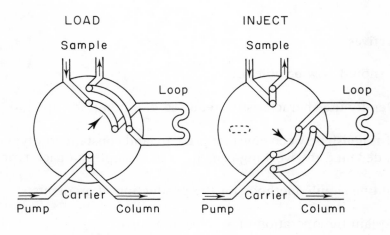

LOAD   INJECT

Sample   Sample

Loop   Loop

Carrier   Carrier

Pump   Column   Pump   Column

(i) External loop valve

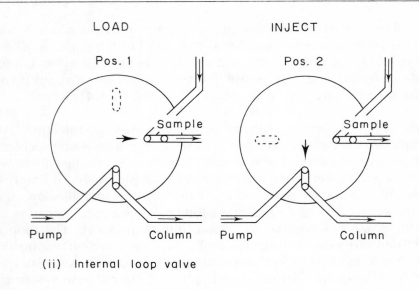

(ii) Internal loop valve

Fig. 2.2g.   *Injection valves*

Summary

The operating principles of three types of hplc pump are described, together with their advantages and limitations. Techniques for the production of solvent gradients and for the introduction of samples are considered.

Objectives

You should now be able to:

- identify the characteristics required of a pump for hplc;

- distinguish between constant pressure and constant flow types and understand the working principles of examples of each type;

- outline simple methods for the production of gradients;

- explain the operation of an injection valve.

2.3. COLUMNS

2.3.1. Column Dimensions and Fittings

The columns most commonly used at the moment are made from stainless steel tube with a 6.35 mm (1/4 inch) external diameter, a 4.6 mm internal diameter and up to 25 cm long. Most manufacturers offer a number of alternative lengths and diameters, eg lengths of 10, 12.5 or 15 cm and internal diameters of 3, 6.2 or 9 mm. The columns can be packed with 10, 5 or 3 μm diameter particles.

At the end of the column there is a stainless steel gauze or frit to retain the packing, and then, for the 4.6 mm type, a 1/4–1/16 stainless steel reducing union with a short length of 0.25 mm (0.01 inch) id ptfe tube to connect the column to the detector (I hope you don't find the inches and mm too confusing; manufacturers tend to give external diameters and fitting sizes in inches, internal diameters in mm and column lengths in cm!).

Conventional reducing unions have rather a large dead volume, so they are bored out to produce the zero dead volume (zdv) union in which both the metal column and the ptfe tube are butted up directly against the stainless steel frit. There is some evidence that the abrupt changes in diameter in the zdv fitting where the two tubes meet can cause some loss of efficiency, so many columns use the more expensive low dead volume (ldv) fitting in which the gauze or frit at the end of the column is followed by a shallow distributive cone leading to the ptfe tube. The dead volume of the ldv type is very small, often about 0.1 μl. The three types are shown in Fig. 2.3a.

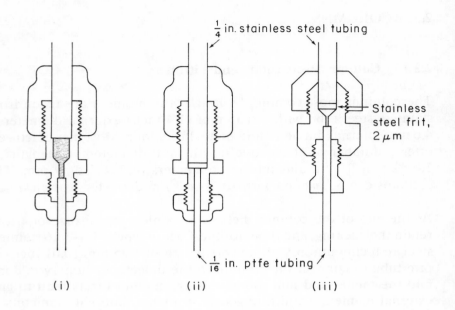

Fig. 2.3a. *Column outlet fittings*

(*i*)   Conventional reducing union (dead volume is shaded)
(*ii*)   zdv union
(*iii*) ldv union

The 4.6 mm column is connected to an injection valve using a zdv or ldv reducing union and a short length of 1/16 inch stainless steel tube. Usually a stainless steel frit is used at the top of the column as well.

Columns are sold by manufactures with a 1/4–1/16 zdv or ldv union at the outlet and a 1/4 inch nut and cap or a reducing union at the inlet. The column will have been tested, and a test chromatogram will be supplied with it. The test chromatogram is, of course, designed to show the column in the best possible light, and the column may not behave as well in another instrument or with another sample.

There are several reasons why hplc columns typically have diameters of around 5 mm. One is that the behaviour of the column is influenced by the dead volume in the rest of the equipment. There are a number of potential advantages in reducing column diameter, but to get good performance from a small bore column the dead volume in the rest of the system has to be very small, and it is only recently that equipment has become available that allows small bore columns to be used properly. To see what the problems are, we will have to look briefly at the way in which the performance of a column is measured, and at some of the factors that cause band spreading or dispersion in chromatography.

2.3.2. Column Performance

One of the difficulties with any form of chromatography is that a band of solute is dispersed, becoming less concentrated as it travels through the system. The efficiency of the column is a measure of the amount of spreading that occurs. In the chromatogram in Fig. 2.3b, V_R = the retention volume of a solute and w_B = the volume occupied by the solute. This is called the peak width, but remember it means a volume, not a length.

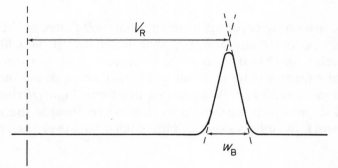

Injection point

Fig. 2.3b. *Measurement of column performance*

The more efficient the column, the smaller will be w_B at a given value of V_R. To measure efficiency, we use quantities called the plate number (N) or the plate height (H) of the column, which are defined as follows:

$$N = 16(V_R/w_B)^2 \qquad (2.3a)$$

$$H = L/N \qquad (2.3b)$$

where L = length of the column

As the column becomes more efficient, N gets larger and H gets smaller. The plate number is dimensionless, but H has units of length, and is usually measured in mm or μm. There are other ways in which N can be measured, depending on whereabouts on the peak we choose to measure its width. Another way is:

$$N = 5.54(V_R/w_{\frac{1}{2}})^2 \qquad (2.3c)$$

where $w_{\frac{1}{2}}$ = width of the peak at half height.

Manufactured hplc columns have about 50 000 plates m^{-1} if packed with 5 μm particles and about 25 000 plates m^{-1} if packed with 10 μm particles, so that from a 12.5 cm column with a 5 μm packing we would expect a plate number of about 6500, corresponding to a plate height of 0.02 mm. Whether or not these high efficiencies are required depends on the sort of work you are doing; a great deal of routine work in hplc is done at efficiencies far lower than this.

2.3.3.   Extra-column Dispersion

Dispersion can be produced outside the column by dead volume in the injector, the detector or the plumbing. The combined effect of all these is called *extra-column dispersion*. Fig. 2.3c shows an example of this, in which different dead volumes are connected between the column and the detector, and Fig. 5.3b shows dispersion produced by dead volume at the top of the column. You can see from these that dead volume effects can cause a serious loss of performance.

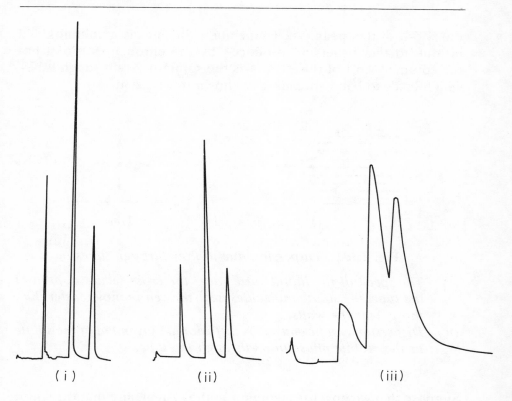

Fig. 2.3c. *Extra-column dispersion*

(i) Normal chromatogram
(ii) Chromatogram obtained with a 15 cm length of 0.8 mm id tube inserted between column and detector inlet tube.
(iii) Chromatogram obtained with a 12.5 cm × 4.6 mm tube inserted as above.

(Sample and conditions were as the example in Section 4.4)

We can examine the effect of dead volume by supposing that w_B in Fig. 2.3b is due to the dispersion produced by the column alone, ie what we would observe if the column was used in an ideal chromatograph that had no dead volume. Now imagine a real chromatograph without a column, in which the injector is connected to the column outlet tube. If we injected a solute we would observe a peak, and the width of this peak, w_A (remember, this means a volume), will be due to the dispersion produced by the chromatograph minus the column. Most of this occurs as the solute passes through tubes, which leads to band spreading as shown in Fig. 2.3d.

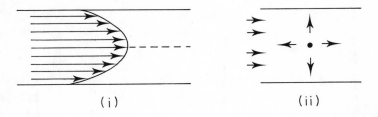

Fig. 2.3d.   *Dispersion due to flow through tubes*

(i)   *The speed of the liquid varies over the cross-sectional area of the tube, the solute molecules near the centre move faster than those near the walls.*

(ii) *Dispersion can be caused by diffusional movement of the solute as the mobile phase flows through the tube.*

Suppose that we now use a column in the system, and that the width of our solute peak is w_T. This will be larger than w_B, because of the extra-column dispersion. The relation between the three peak widths is:

$$w_T^2 = w_B^2 + w_A^2 \qquad (2.3d)$$

The value of w_A is likely to be 40–60 μl, unless the system has been specially designed for use with small bore columns. The effect on a chromatogram is more serious the earlier a peak is eluted, so we will examine the worst case, which is the effect on an unretained solute (one that travels at the same speed as the mobile phase).

∏ In a column packed with 5 μm silica, the solvent occupies about 70% of the volume of the column. For a 25 cm × 4.6 mm column packed with 5 μm silica, calculate:

(a) The retention volume (in μl) of an unretained solute

(b) The width w_B (in μl) of the unretained solute peak. Use Eq. 2.3a and assume that the plate number of the column is 10 000.

(c) The width w_T that we would observe for this peak. Use Eq. 2.3d and assume that the extra-column dispersion is 50 μl.

Remember that 1 cm^3 = 1000 mm^3 = 1000 μl.

(a) The retention volume of an unretained solute is equal to the volume of solvent in the column.

$$\text{volume of empty column} = \pi \times (4.6)^2 \times 250/4$$

$$= 4155 \; \mu l$$

$$\text{volume of solvent in the column} = 4155 \times 0.7$$

$$= 2908.5 \; \mu l$$

(b) $10000 = 16 \times (2908.5/w_B)^2$

∴ $w_B = 116.3 \; \mu l$

(c) $w_T^2 = 116.3^2 + 50^2$

∴ $w_T = 126 \; \mu l.$

The extra-column dispersion governs the dimensions of the column that we use. In the calculation above, the dispersion is increased by about 8% by the extra-column effects. If we want the dispersion to be increased by no more than this, then w_B should not be any smaller than the value calculated above. This in turn limits the retention volume, and thus the volume of the column itself. The minimum column volume we can use will depend on the amount of extra-column dispersion and on what we consider to be an acceptable increase in peak width that is produced by extra-column effects. In practice, this acceptable increase is taken as 10%, based on an unretained solute, and if we take 50 μl as a typical figure for extra-column dispersion then the minimum column diameter works out at about 4.5 mm for a column 25 cm long.

SAQ 2.3a

In Section 2.3.3 you worked out the effect of extra column dispersion on the peak of an unretained solute, using a column with a plate number of 10 000. The extra-column dispersion will decrease the plate number that we actually observe for this column. The table below contains the retention volume, peak widths and plate number for an unretained solute on this column.

(*a*) Complete the table by calculating the corresponding values for columns with plate numbers of 5000 and 3000 respectively.

(*b*) What is the % reduction in the plate number of each column due to extra-column effects. $\longrightarrow$

**SAQ 2.3a
(cont.)**

N (ideally)	10 000	5000	3000
V_R, μl	2908.5	2908.5	2908.5
w_B, μl	116		
w_T, μl	126		
N (actual)	8525		
% reduction in N			

2.3.4. Column Dispersion Mechanisms

There are several mechanisms responsible for the dispersion of solute as it travels through the column:

(*a*) Multiple path effect (eddy diffusion) and lateral diffusion (flow dispersion). Multiple path is the term for the dispersion produced by the existence of different flow paths, by which solute species can progress through the column. These path differences arise because the stationary phase particles may have irregular shapes, and also because the packing of the column may be imperfect. The smaller the particles and the narrower their size distribution the less the dispersion. If solute species travel at the same speed, those in different flow paths will travel different distances in the column in a given time. In fact, those in different flow paths will travel at different speeds, moving faster in the wider flow paths and slower in the narrower ones. Even in the same flow path, the solute species will move at different speeds, those in midstream travelling faster than those close to the stationary phase particles (See Fig. 2.3d).

It is also possible for solute species to diffuse laterally (in a radial direction) across the column, and thus to move from one flow path into another. This effect reduces the amount of dispersion produced by the multiple path effect as it tends to equalise the speed of the solute species in the column. The longer the time a solute species spends in the column, the more lateral diffusion will occur, so that flow dispersion is reduced by using low flow rates of mobile phase.

(*b*) Longitudinal diffusion. Dispersion also arises because of diffusion of solute species in the longitudinal (axial) direction in the column.

∏   This is an important source of dispersion in gas chromatography, less so in practice in liquid chromatography. Why do you think this is so?

Rates of diffusion are very much slower (roughly 10^4 to 10^5 times slower) in liquids than in gases.

Longitudinal diffusion will become more serious the longer the solute species spend in the column, so this effect, unlike flow dispersion is reduced by using a rapid flow rate of mobile phase.

(*c*) Mass transfer effects. These effects arise because the rate of the distribution process (sorption and desorption) of the solute species between mobile and stationary phases may be slow compared to the rate at which the solute is moving in the mobile phase. When solute species interact with the stationary phase they may spend some time in or on the stationary phase before rejoining the mobile phase, and in this time they will have been left behind by those solute species which did not interact. The internal pores of the stationary phase particles will contain 'stagnant' mobile phase, through which the solute species have to diffuse before they can get at the stationary phase. Those that diffuse a long way into the porous structure will be left behind by those species that bypass the particle, or only diffuse a short distance into it.

∏   Do you think these mass transfer effects would become more serious at low or high mobile phase flow rates?

In the time it takes for mass transfer into or out of the stationary phase, we want a non-interacting solute species to have travelled as little as possible in the column, so we need a low flow rate.

The three effects are shown in Fig. 2.3d and 2.3e.

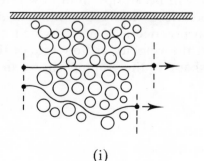

(i)

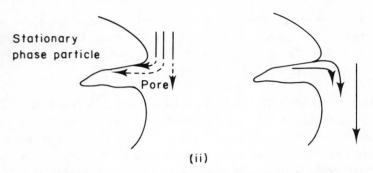

(ii)

Fig. 2.3e.   *Column dispersion mechanisms*

(i) Multiple path effect (flow dispersion)

> *Two molecules start at the same place in the column. If they both travel at the same speed, the molecule with the simpler flow path travels further in the column in a given time. Flow path differences may be greater in the wall regions of the column, where packing is irregular.*

(ii) Mass transfer

> *Solute molecules which diffuse into the stationary phase particles and interact with them are left behind by those molecules that bypass the stationary phase.*

You can see that these dispersion mechanisms are affected in different ways by the flow rate of mobile phase. To reduce dispersion due to longitudinal diffusion we need a high flow rate, whereas a low flow rate is needed to reduce dispersion due to the other two. This suggests that there will be an optimum flow rate where the combination of the three effects produces minimum dispersion, and this can be observed in practice if N or H (which measure dispersion) are plotted against the velocity or flow rate of the mobile phase in the column. The shape of the graph is shown in Fig. 2.3f.

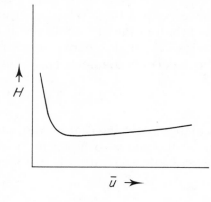

Fig. 2.3f. *Plate height versus mobile phase velocity*

The total column dispersion is due to the combined effects of flow dispersion, longitudinal diffusion and mass transfer.

As the mobile phase velocity increases:

Flow dispersion increases
Dispersion due to slow mass transfer increases
Dispersion due to longitudinal diffusion decreases.

In practice, because of the slow diffusion rates in liquids, dispersion due to longitudinal diffusion becomes important only at very low velocities. Because the dispersion increases only slowly with increasing mobile phase velocity, flow rates used in hplc are considerably higher than the value corresponding to minimum dispersion. This gives us fast separations without too much loss of efficiency.

Irregularities in the column packing can often occur where the packing meets the wall of the column. Such irregularities accentuate flow path differences, thus producing dispersion. The nature and extent of the irregularities will depend on the roughness of the wall and how well the column has been packed, and is reduced by internal polishing of stainless steel columns or by the use of other materials, such as glass or plastics, to form the column wall. Ideally, we do not want our solutes to travel in the wall regions of the column. Fortunately, in practice radial dispersion of solutes in hplc is quite slow, and it has been shown that columns can be operated under conditions such that solutes never reach the wall regions at all as they travel down the column. When operated in this way, a column is called an 'infinite diameter' column. For infinite diameter behaviour, the following relation between column diameter (d_c), column length (L) and particle diameter (d_p) must be obeyed:

$$d_c^2/(d_p \times L) \geqslant 2.4 \qquad (2.3e)$$

This relation is calculated assuming that no more than 5% of solute reaches the column walls, and also assuming a point injection of solute (ie the volume of injected solute is negligible) at the centre of column.

∏   Given these conditions, what would be the minimum diameter of a 25 cm column, packed with 5 μm particles, if the column is to show infinite diameter behaviour?

Working in mm, $d_c^2 = 2.4 \times 5 \times 10^{-3} \times 250$

$\therefore \quad d_c = 1.7$ mm.

In practice, the injected solute will occupy a finite volume, and the injection may not be properly centralised either. Both of these have the effect of making d_c greater, or L smaller, for infinite diameter behaviour, than the values calculated from Eq. 2.3e. If these effects are allowed for, it can be shown that columns of 25 cm or less, with injections of 10 μl or less, show infinite diameter behaviour when their diameters are greater then about 4 mm.

2.3.5. Other Types of Column

A number of manufacturers (eg Chrompack, Merck, Waters) produce columns made of glass, glass lined stainless steel or plastic that use a 'cartridge' system in which the column is supplied as a cartridge and the end fittings are contained in a separate cartridge holder. These columns are much easier to change than conventional columns that are terminated with compression fittings. Fixing the column in the holder in some systems is done simply by operating a compression lever which locks the column leak tight into the holder.

Because they are sold without end fittings, cartridge-type columns are cheaper than conventional columns, although the initial outlay on the cartridge holder has to be considered.

The Waters system uses a plastic cartridge which is inserted into a device (the Z-module) that subjects the column to radial compression, ie pressure is applied along the radial axis of the column tube. The flexible wall of the column then moulds itself into the voids that are present in the wall regions of the column. This method is claimed to produce an improvement in the packed bed structure, better column performance and longer useful column life.

Although most workers in hplc prefer to use manufactured columns, it is not difficult to make your own. This is dealt with in Section 5.1.

2.3.6. Small Bore Columns

Small bore or microbore is a term used for hplc columns that have diameters less than about 2 mm. Columns of this type were first used as long ago as 1967, but at that time the influence of extra-column dispersion was not appreciated, so that the columns were not used in chromatographs of appropriate design. In 1977 there was a renewal of interest in the properties of small bore columns, but it is only in the last few years that systems have become commercially available that allow the potential of small bore columns to be realised. Several manufacturers are now marketing a range of small bore columns, and a number of recent hplc instruments are claimed to be compatible with them.

One advantage of small bore columns is that they are operated at much lower mobile phase flow rates than the 4.6 mm columns, so there is a large reduction in solvent consumption and hence operating costs of the chromatograph. The efficiency of hplc columns does not depend on their diameter but it does depend on the velocity of the mobile phase in the column, so microbore columns are operated at velocities corresponding to the flow rates used with larger columns. If f (cm^3 min^{-1}) is the flow rate in a column with diameter d cm, and the mobile phase velocity is v cm min^{-1}, f and v are related by:

$$f = \pi d^2 v/4 \tag{2.3f}$$

Also, if column 1 with diameter d_1 is operated at f_1 cm^3 min^{-1} and we want to operate column 2 which has a diameter d_2 at the same mobile phase velocity, then the flow rate f_2 that we need is given by:

$$f_2 = f_1 \times d_2^2/d_1^2 \tag{2.3g}$$

∏   These relations expressed by Eqs. 2.3f and 2.3g are both easy to prove; see if you can do it.

In 1 minute the mobile phase travels v cm along the column. The volume of mobile phase contained in a cylinder of length v and diameter d will be $\pi d^2 v/4$. This will be numerically equal to the flow rate.

For columns 1 and 2 we have $f_1 = \pi d_1^2 v/4$ and $f_2 = \pi d_2^2 v/4$; dividing these produces Eq. 2.3g.

∏   Use Eq. 2.3g to calculate the missing data in Fig. 2.3g. The correct figures are given at the end of the Section. An example of the money that can be saved by the use of these low flow rates is given in SAQ 2.3b.

Flow rate in 4.6 mm column, cm^3 min^{-1}	Flow rate in small bore column at the same mobile phase velocity, μl min^{-1}	
	column diameter, mm	
	2	1
1		
2		
5	945	

Fig. 2.3g.   *To be completed*

Another advantage of small bore columns concerns the small peak widths that are produced. In Section 2.3.3 we calculated the retention volume and the peak width of an unretained peak from a 25 cm × 4.6 mm column.

∏   See if you can repeat the calculation for a 25 cm column with a diameter of 1 mm. Assume as before that the plate number of the column is 10 000 and that the solvent occupies 70% of the volume of the column. Put the results in Fig. 2.3h (the correct figures are given at the end of the Section).

| | column diameter, mm | |
	4.6	1
Volume of column, μl	4155	
V_R, μl	2909	
w_B, μl	116	

Fig. 2.3h.   *To be completed*

Comparing the small and the large bore columns, if we injected the same mass of solute each time, then for the small bore column the same mass of solute would be present in a smaller peak volume, which means that we can get the same concentration of solute in a peak from a small bore column using a smaller mass of solute. The detector in our system will respond to the concentration of solute, so that with the small bore columns we will be able to detect smaller amounts of material. This is important in dealing with samples where only small quantities are available, for instance those of biological origin.

Small bore columns require equipment with very low extra-column dispersion, and this is the main difficulty associated with their use.

∏   For the 1 mm column in Fig. 2.3h:

(a) Calculate the extra-column dispersion we need if it is only to increase the column dispersion by 10%.

(b) What would be the plate number for an unretained peak on this column if we used the column in a conventional chromatograph with an extra-column dispersion of 50 μl?

(a) $w_T = 5.5 + 0.55 = 6.05 \ \mu l$

$6.05^2 = 5.5^2 + w_A^2$

$\therefore \quad w_A = 2.5 \ \mu l$

(b) In a chromatograph with 50 μl dispersion

$w_T^2 = 5.5^2 + 50^2$

$\therefore \quad w_T = 50.3 \ \mu l$

The plate number for an unretained peak will be

$$N = 16 \times 137^2/50.3^2 = 119$$

thus it would be pointless to use the column in a conventional system, as the performance of the column would be completely lost!

A value of 2.5 μl for extra-column dispersion does not tell us what our system dead volume needs to be (except that it needs to be very small). Much of the extra-column dispersion in a chromatograph can be considered to occur as the solute passes through tubes, as shown in Fig. 2.3d. This effect is well understood, so that the dispersion produced can be calculated in some cases if the various dimensions are known.

The volume of solute that is injected causes some dispersion, so that for small bore columns this has to be low, usually less than

1 μl. There are several (expensive) small volume valve injectors that will do; the lowest volume available at the moment is 0.06 μl. It is also possible to use an external loop injector and switch it to the column for a short time only, so that the full volume of the sample loop is not delivered. The internal volume of detectors has to be reduced without causing loss of sensitivity. There are a few detectors on the market at present with small enough cell volumes; other types can sometimes be modified. Electrochemical detectors (Section 2.4.4) are the easiest types to make with a small internal volume, so that these will probably be useful for detecting electroactive solutes separated on small bore columns.

Most of the hplc instrumentation now in use is unsuitable for small bore columns. At the moment, the technique is used mainly in the applications laboratories of some instrument manufacturers (they are interested in selling it!). The method is potentially attractive in areas where sample sizes are very limited, for example in biochemical or life sciences applications, but whether or not it becomes widely accepted remains to be seen.

Flow rate in 4.6 mm column, cm^3 min^{-1}	Flow rate in small bore column at the same solvent velocity, μl min^{-1}	
	column diameter, mm	
	2	1
1	189	47
2	378	95
5	945	236

Completed Fig. 2.3g

	column diameter, mm	
	4.6	1
Volume of column, μl	4155	196
V_R, μl	2909	137
w_B, μl	116	5.5

Completed Fig. 2.3h

SAQ 2.3b

Suppose you are running a 4.6 mm hplc column on a mixture of acetonitrile and water (80% by volume acetonitrile). The column runs continuously for 8 hours a day at a flow rate of 2 cm^3 min^{-1}, and your acetonitrile costs you seven pound per dm^3.

(a) What is the cost of acetonitrile per year, assuming a year = 250 working days?

(b) What is the mobile phase velocity (cm min^{-1}) through the column?

(c) If you changed to a 1 mm column operated at the same velocity, what flow rate would you have to use?

(d) What would the small bore column save you in acetonitrile costs?

SAQ 2.3b

2.3.7. Fast Separations Using 3 × 3 Columns

A typical hplc column (25 cm × 4.6 mm, packed with a 5 μm bonded silica stationary phase) will have an efficiency corresponding to a plate number of 10 000–15 000. For many separations, this efficiency is far more than is needed, as often a plate number of 3000–5000 will give baseline resolution of all solutes. If this is the case, using a conventional column will waste analysis time and solvent.

Short (3.3 cm × 4.6 mm) columns packed with 3 μm bonded silica stationary phases have sufficient efficiency for many separations. They are commonly called 3 × 3 columns, and, compared to conventional columns, have the following advantages:

(*a*) They perform a given separation in a shorter time, with the use of less mobile phase

(*b*) They equilibrate rapidly with the mobile phase and are thus useful for methods development

(*c*) Solutes are eluted with small peak volumes giving a greater mass sensitivity, an advantage when the amount of sample is limited

(*d*) They are slightly cheaper.

You may have noticed that some of these are the same sort of advantages that are claimed for microbore columns.

∏   What precautions do you think would be necessary when using 3 × 3 columns?

Because of their small internal volume, they require equipment with low extra-column dispersion. This is not, however as serious a limitation as it is with microbore columns, and 3 × 3 columns can often be used satisfactorily in conventional instruments.

The small particle size of the packing makes 3 × 3 columns susceptible to plugging problems. Samples and mobile phases should be filtered before use, and the column should be protected with guard and scavenger columns (see Section 5.3.2).

Fig. 2.3i shows the separation of a mixture of six explosives, both on a 3 × 3 column and on a conventional column. The same quantity of sample was used for each chromatogram. On the 3 × 3 column a faster separation is obtained, with a higher mass sensitivity.

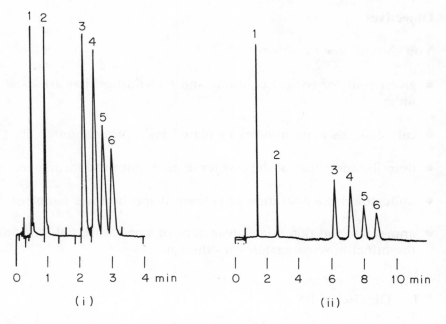

Fig. 2.3i. *Separation of polynitro explosives*

Column: 5 μm C-8 bonded phase (*i*) 3.3 cm × 4.6 mm
                                                     (*ii*) 15 cm × 4.6 mm
Mobile phase: tetrahydrofuran/methanol/water 2 : 29 : 69
Flow rate: (*i*) 2 cm^3 min^{-1} (*ii*) 3 cm^3 min^{-1}
Detector: Uv absorption, 230 nm
Injection: 1 μl CH$_3$CN containing 100 ng of each solute
Sample: 1 = HMX 2 = RDX 3 = 2,4-DNT 4 = Tetryl 5 = TNT
                 6 = 2,6-DNT

Summary

Typical dimensions and performance figures are given for conventional hplc columns. The effect of dead volume on column performance is considered, and other dispersion mechanisms are described. The advantages and limitations of small bore columns are discussed.

Objectives

You should now be able to:

● give details of typical columns and end fittings that are used in hplc;

● calculate the plate number or plate height of a column;

● describe the effect of dead volume on column performance;

● understand the operation of column dispersion mechanisms;

● appreciate the potential advantages of small bore columns, and the difficulties associated with their use.

2.4.   DETECTORS

2.4.1.   Introduction

The function of the detector in hplc is to monitor the mobile phase emerging from the column. The output of the detector is an electrical signal that is proportional to some property of the mobile phase and/or the solutes. Refractive index, for example, is a property of both the solutes and the mobile phase. A detector that measures such a property is called a *bulk property* detector. Alternatively, if the property is possessed essentially by the solute, such as absorption of uv/visible radiation or electrochemical activity, the detector is called a *solute property* detector. Quite a large number of devices, some of them rather complicated and tempremental, have been used as hplc detectors, but only a few have become generally useful, and we will examine five such types. Before doing this, it is helpful to have an idea of the sort of characteristics that are required of a detector.

∏   What do you think they are? (there are about seven important ones).

The important characteristics are:

(*a*) sensitivity
(*b*) linearity
(*c*) universal or selective response
(*d*) predictable response, unaffected by changes in conditions
(*e*) low dead volume
(*f*) nondestructive
(*g*) cheap, reliable and easy to use.

No detector possesses all of these, and unfortunately the properties are not all independent of one another, so that improving a detector in one respect may make it worse in another.

(*a*) Sensitivity is the ratio of output to input, so that we want a large detector signal for a small amount of solute. Any detector will suffer from instrumental, principally electronic, noise, the amplitude of which will determine the minimum amount of solute that we can detect, because as we decrease the amount of our solute, eventually the detector signal will become so small that it cannot be distinguished from the noise. The sensitivity of detectors is often given as a noise equivalent concentration, C_N, which means the concentration of solute that produces a signal equal to the detector noise level. C_N will depend on the nature of the solute that is used to make the test.

(*b*) A linear detector has a response that is directly proportional to the amount or concentration of solute. The linear range of the detector is that concentration range over which this proportionality is obeyed. It is still possible to use a detector if it is not linear, but quantitative work becomes much more difficult.

(*c*) and (*d*) A universal detector will sense everything in the sample, whereas a selective detector will sense only certain components. In analytical work we need both types. Ideally, our detector would have the same very high sensitivity for all solutes, be capable of operating universally or selectively and have a response that did not depend on the operating conditions, but this is asking far too much of it. The best we can hope for is that we can predict how the response of the detector will change for different chemical types, and that the response does not change too much if there are small changes in the operating conditions (eg column temperature or flow rate).

(*e*) Dead volume in the detector adds to extra-column dispersion, so it must be kept to a minimum. This includes the cell volume of the detector itself, and also the length and bore of any tubing associated with it. For spectrometric detectors a reduction in the cell volume is likely to lead to a loss of sensitivity.

Some of these characteristics are listed for different detectors in Fig. 2.4a.

Type	Response	Noise level	C_N g cm^{-3}	Linear range	Flow cell volume, μl
Uv-visible absorption	selective	10^{-4} au	10^{-8}	10^4–10^5	1–8
Fluorescence	selective	10^{-7} au	10^{-12}	10^3–10^4	8–25
Conductivity	selective	10^{-2} μS cm^{-1}	10^{-7}	10^3–10^4	1–5
Amperometric	selective	0.1 nA	10^{-10}	10^4–10^5	0.5–5
Refractive index	universal*	10^{-7} riu	10^{-6}	10^3–10^4	5–15

* There must be a difference between the refractive index of the solutes and that of the mobile phase.

Fig. 2.4a.   *Characteristics of detectors used in hplc*

Noise levels will be different for different models of the same type of detector, and for a given model will depend very much on how the detector is used. The noise equivalent concentration refers to a solute with favourable properties, and may be very much higher for other solutes.

2.4.2. Uv Absorbance Detectors

These are by far the most popular detectors in hplc. The principle is that the mobile phase from the column is passed through a small flow cell held in the radiation beam of a uv/visible photometer or spectrophotometer. These detectors are selective in the sense that they will detect only those solutes that absorb uv (or visible) radiation. Such solutes include alkenes, aromatics and compounds having multiple bonds between C and O, N or S. The mobile phase we use, on the other hand, should absorb little or no radiation.

Absorption of radiation by solutes as a function of concentration, c, is described by the Beer–Lambert law:

$$A = \epsilon . c . b \qquad (2.4a)$$

Where A = absorbance, b = path length of the cell and ϵ = molar absorptivity, which is a constant for a given solute and wavelength.

The magnitude and the units of the absorptivity in Eq. 2.4a will depend on the units of c and b. In SI units, with c in mol m^{-3} and b in m, ϵ is in mol^{-1} m^2, but it is common practice to measure c in mol dm^{-3} and b in cm, when ϵ will be in dm^3 mol^{-1} cm^{-1}.

Strictly, the Beer–Lambert law applies only to monochromatic radiation. However, the detector system does not provide truly monochromatic radiation, but rather a narrow band of wavelengths centered around the selected wavelength. If we test the law for a solute at a wavelength in the spectrum where the absorbance is changing rapidly with wavelength, then the different wavelengths comprising the band may be absorbed by quite different amounts and the law may not be obeyed. We want to operate our detector in flat regions of the spectrum (maxima, minima or shoulders). Whether or not the law is obeyed in steep regions of the spectrum depends on the quality of the detector.

Both fixed and variable wavelength uv/visible detectors are available. The variable types use a deuterium and/or a tungsten filament lamp as the radiation source and can operate between about 190–700 nm. They will have a number of absorbance ranges (ranges are given in 'aufs' which means absorbance units corresponding to full scale deflection on the recorder). Fixed wavelength detectors can operate at 254 nm, 280 nm or at other wavelengths. Fig. 2.4b shows some of the specification of two modern variable wavelength detectors.

	Radiation source	wavelength range, nm	absorbance ranges, aufs	noise, au
Phillips 4025	Deuterium lamp	190–380	0.005–1.28 (9 ranges)	10^{-4} at 230 nm
	Tungsten lamp (accessory)	190–600		
Uvikon 735	Deuterium lamp	195–350	0.0025–2.56 (11 ranges)	5×10^{-4} at 250 nm

Fig. 2.4b. *Specification of two variable wavelength detectors*

∏   Suppose you are using a uv detector with a noise level of 10^{-4} au. The detector is linear from the noise level to A = 1, and a flow cell with a path length of 10 mm is used.

(*a*) What is the linear range of the detector?

(*b*) Use Eq. 2.4a to calculate C_N (in g cm^{-3}) for a solute with $M_r = 100$ and absorptivity = 1000 dm^3 mol^{-1} cm^{-1}.

(*a*) 10^4

(*b*) If c = the concentration of solute that produces an absorbance of 10^{-4}, then $10^{-4} = 1000 \times 1 \times c$

$$\therefore \quad c = 10^{-7} \text{ mol dm}^{-3}$$
$$= 10^{-5} \text{ g dm}^{-3}$$
$$= 10^{-8} \text{ g cm}^{-3}.$$

This would be lower for another compound that had a higher absorptivity.

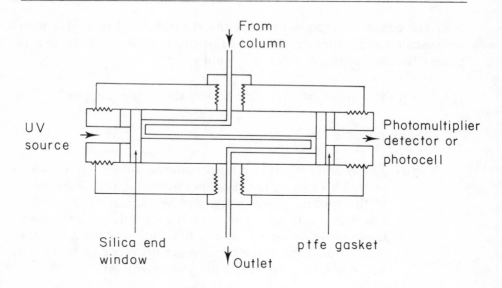

Fig. 2.4c. *Flow cell for uv-visible absorbance detector*

Fig. 2.4c is a diagram of a simple type of uv flow cell. The cell has a 1 mm internal diameter and the optical path is 10 mm giving it an internal volume of just under 8 μl. Modern instruments use a 'cassette' type flow cell, which plugs into a holder in the detector; in older detectors there are arrangements for horizontal and vertical adjustment of the position of the flow cell in the radiation beam. More complicated flow cell designs attempt to reduce flow disturbances, which can be caused by changes in the refractive index of the eluent, for instance if the solute has been dissolved in a solvent other than the mobile phase. Fig. 2.3d shows that the material will appear first in the centre of the flow cell and last at the walls, forming what amounts to a moving lens of liquid in the cell. This distorts the radiation beam, and may either increase or decrease the amount of radiation falling on the sensor in the instrument. Such changes are often seen as a differential peak on the chromatogram at about the time expected for an unretained solute.

There are several other problems that can occur with uv flow cells. Air bubbles in the cell can produce a series of very fast noise spikes on the chromatogram, or pronounced drift in the baseline, followed by sudden changes in the baseline position, or both effects at once.

Another common problem is that the detector registers very high or offscale absorbance readings all the time, ie the uv radiation is being absorbed strongly when it should not be.

∏   Can you think of any reasons why this might happen?

The possible causes are:

(a) The mobile phase contains some uv absorbing component. This can be checked by measuring the absorbance of the mobile phase using another spectrophotometer, but make sure that you take the sample directly from the mobile phase reservoir. It is not unknown for the mobile phase to have been made up incorrectly, and if you make up a fresh sample you might get it right!

(b) There are large air bubbles in the flow cell. These can sometimes be removed by pumping at a high flow rate or by disconnecting the detector from the column and passing solvent rapidly through the flow cell using a syringe. This can be done with a syringe of 10–20 cm^3 capacity with a 1/16 union fitted to the end of the needle.

(c) The flow cell may be leaking so that there are drops of solvent on the outside of the end windows, or the end windows may be dirty, or the cell may not be properly aligned in the instrument. The alignment is easily checked, but only dismantle and clean the flow cell as a last resort; some types are quite difficult to reassemble. Faults in the detector can be checked by seeing if you can obtain zero absorbance with the flow cell removed; if you can, the detector is probably alright.

In uv absorbance detection, it is often useful to be able to detect different peaks in the chromatogram at different wavelengths. This may be because certain solutes have only a small absorptivity at the wavelength selected. At other times, a wavelength change can be used to 'edit out' unwanted peaks. It is also useful to be able to record the spectrum of each component; this can tell us the optimum detection wavelength or combination of wavelengths and it can sometimes be used to identify the peaks.

In a *multichannel* or *diode array* detector, polychromatic radiation is passed through the flow cell and the emerging radiation is diffracted by a grating so that it falls on to an array of photodiodes. Each photodiode receives a different narrow wavelength band. The complete array of diodes is scanned by a microprocessor many times a second. The resulting spectra may be displayed on the screen of a vdu and/or stored in the instrument for subsequent transfer to hard copy, eg a recorder or printer, at the end of the chromatographic run. If a suitable data station is used in conjunction with the detector you can do a variety of post-run tasks, such as comparison of the spectra with a library of standard spectra recalled from disc storage. The multichannel detector can provide detection at a single wavelength, or at a number of wavelengths simultaneously, or detection wavelength changes can be programmed to occur at specified times in the chromatogram. Absorbance ratios at selected wavelengths can also be displayed for each peak. To see the advantage of doing this, look at the spectrum in Fig. 2.4d.

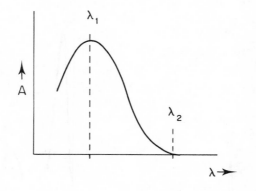

Fig. 2.4d. *Absorbance ratios*

The absorbance ratio $A_{\lambda2}/A_{\lambda1}$ for the solute peak should be close to zero. If it is not, then this suggests that the peak is not what we think it is. For example, there may be another component that elutes at the same time, so the ratio method is a simple way of indicating the purity of the peaks.

Fig. 2.4e shows how a multichannel detector can be used to clean up a chromatogram by using programmed changes in the detection wavelength. In this example, three anticonvulsant drugs are determined in a blood sample. In (i), with detection at a single wavelength, two of the peaks are partly obscured by excipients (this means, all the assorted rubbish in the sample that we are not interested in). From the spectra, (ii)–(iv), we can find a wavelength that gives acceptable sensitivity for peak 1 and discriminates against the interfering peaks. With programmed wavelength detection, (v), 274 nm is used until just before peak 2 elutes, then the wavelength is changed to 210 nm.

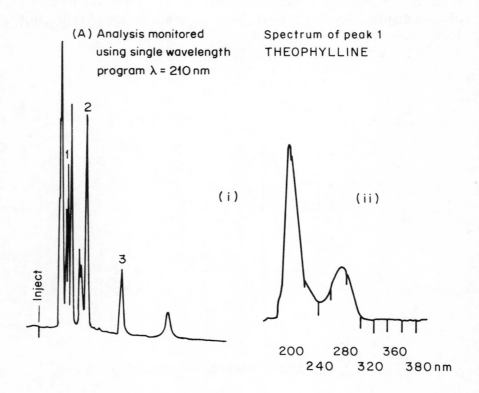

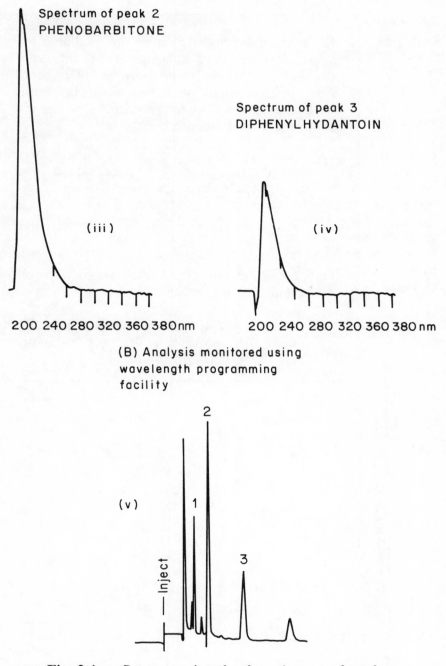

Spectrum of peak 2
PHENOBARBITONE

Spectrum of peak 3
DIPHENYLHYDANTOIN

(iii)

(iv)

200 240 280 320 360 380 nm

200 240 280 320 360 380 nm

(B) Analysis monitored using
wavelength programming
facility

2

1

3

(v)

Inject

Fig. 2.4e *Programming the detection wavelength*

SAQ 2.4a

Infrared absorption detectors are available for hplc, although they have never become very popular. From what you know about ir spectrometry and what you have read so far about hplc detectors, see if you can decide whether the following statements are true or false.

(*i*)   An infrared spectrum provides more structural information about a compound than does a uv spectrum.

(*ii*)   An ir detector would be more sensitive than a uv detector.

(*iii*) An ir detector could not be used with solvent mixtures containing water.

(*iv*)   An ir detector could be used as a selective detector or a universal detector, by changing the wavelength used.

SAQ 2.4b   Fig. 2.4f shows the uv spectra of azobenzene (Az, concentration 3.73×10^{-3} g dm^{-3}) and phenanthrene (P, 3.23×10^{-3} g dm^{-3}) both recorded in *iso*-octane. The wavelength drive on the instrument was 10 nm cm^{-1} and the absorbance range was 2 aufs. Measurements were made against *iso*-octane using 10 mm cells.

What wavelength would you choose:

(*i*)   To detect Az without detecting P

(*ii*)   To detect P without detecting Az

(*iii*)   To detect both of them

(*iv*)   To detect Az at maximum sensitivity?

Calculate the molar absorptivity of each of them (in dm^3 mol^{-1} cm^{-1}) at the wavelength you chose in (*iv*).

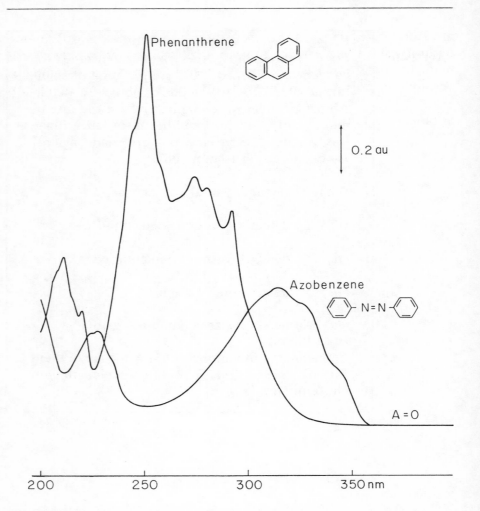

Fig. 2.4f.   *Uv absorption spectra of phenanthrene and azobenzene*

2.4.3.   Fluorescence Detectors

Many compounds are capable of absorbing uv radiation and subsequently emitting radiation of a longer wavelength, either instantly (fluorescence) or after a time delay (phosphorescence). Usually, the fraction of the absorbed energy that is re-emitted is quite low, but for a few compounds values of 0.1–1 are obtained, and such compounds are suitable for fluorescence detection. Compounds that fluoresce naturally have a conjugated cyclic structure (eg polynuclear aromatic hydrocarbons). Many non-fluorescent compounds can be converted to fluorescent derivatives by treatment with suitable reagents.

A block diagram of a fluorescence detector is shown in Fig. 2.4g.

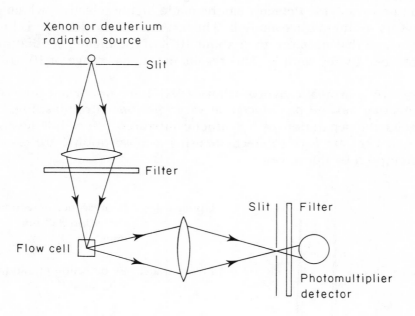

Fig. 2.4g. *Fluorescence detector*

Radiation from a xenon or deuterium source is focussed on the flow cell. An interchangeable filter allows different excitation wavelengths to be used. The fluorescent radiation is emitted by the sample in all directions, but is usually measured at 90° to the incident beam. In some types, to increase sensitivity, the fluorescent radiation is reflected and focussed by a parabolic mirror. The second filter isolates a suitable wavelength from the fluorescence spectrum and prevents any scattered light from the source from reaching the photomultiplier detector. The 90° optics allow monitoring of the incident beam as well, so that dual uv absorption and fluorescence detection is possible, and some commercial models have this facility (Fig. 2.4o is an example).

Fluorescence detectors can be made much more sensitive than uv absorbance detectors; for favourable solutes (such as anthracene) the noise equivalent concentration can be as low as 10^{-12} g cm^{-3}. Because both the excitation wavelength and the detected wavelength can be varied, the detector can be made highly selective, which can be very useful in trace analysis. The response of the detector is linear provided that no more than about 10% of the incident radiation is absorbed by the sample. This results in a linear range of 10^3–10^4.

Polycyclic aromatic hydrocarbons (PAH) are important air pollutants that have to be detected at very low concentrations. Fig. 2.4h shows the separation of a synthetic mixture of very low levels of PAH. They are barely detectable using uv absorption, but are easily monitored by fluorescence.

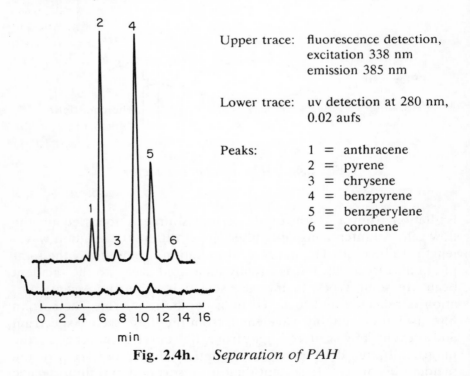

Upper trace: fluorescence detection,
               excitation 338 nm
               emission 385 nm

Lower trace: uv detection at 280 nm,
               0.02 aufs

Peaks: 1 = anthracene
               2 = pyrene
               3 = chrysene
               4 = benzpyrene
               5 = benzperylene
               6 = coronene

Fig. 2.4h.   *Separation of PAH*

2.4.4. Electrochemical (ec) Detectors

Electrochemical detectors measure either the conductance of the eluent, or the current associated with the oxidation or reduction of solutes. To be capable of detection, in the first case the solutes must be ionic and in the second case the solutes must be relatively easy to oxidise or reduce.

The first type are called *conductivity* detectors, although what they measure is the conductance (1/resistance) of the eluent, rather than the conductivity (1/resistivity). Conductivity can be used for the detection of inorganic or organic ions, usually after separation by ion-exchange chromatography. The major problem with this is that ion exchange stationary phases need an ionic mobile phase which itself has quite a high conductance, so to use conductance measurements as a method of detection might at first seem to be unpromising, as we would be measuring small changes in a relatively very large quantity. Because inorganic ions are not easy to detect by other methods, ways have been found to overcome the problem of the conductance of the mobile phase. These are dealt with in Section 3.3.2. Section 2.4.6 describes a combined uv absorbance, fluorescence and conductivity detector.

Electrochemical detectors that measure current associated with the oxidation or reduction of solutes are called *amperometric* or *coulometric* detectors. The term 'ec detector' normally refers to these types rather than conductivity detectors.

When current is passed through a solution, reactions occur at each electrode in which electron exchange takes place between the electrode and substances in solution. At the cathode, substances in solution gain electrons (reduction) and at the anode they lose electrons (oxidation). We can think of the cathode and the anode as a reducing agent and an oxidising agent respectively whose strength depends on the value of the electrode potential. A cathode becomes a stronger reducing agent as its electrode potential becomes more negative and an anode becomes a stronger oxidising agent as its electrode potential becomes more positive. If a substance can be electrochemically oxidised or reduced it is said to be electroactive. If it is difficult to oxidise or reduce chemically, it will be difficult to do so electrochemically as well, so that to reduce it will require a cathode with a relatively large negative potential or to oxidise it will require an anode with a relatively large positive potential. Because we are dealing with a substance in solution, we always have to consider the ec reduction or oxidation of the solvent as well.

In Fig. 2.4i we are plotting the potential E of an electrode (measured against a suitable reference electrode) against the current flowing in the cell. Because oxidation and reduction result in different directions of current flow (for reduction electrons flow out of the electrode and for oxidation electrons flow into it) we distinguish between the two by calling a reduction current (a cathodic current) positive and an oxidation current (an anodic current) negative. The solid line in the figure shows the current-potential curve obtained for a solution containing electroactive solutes A and B which are oxidisable, and C which is reducible. The broken line shows the background curve (which would be obtained without the solutes).

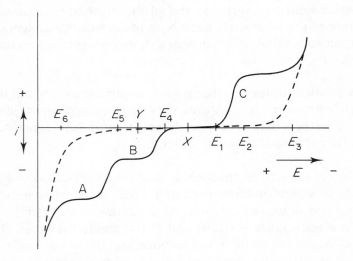

Fig. 2.4i.   *Current-potential curves for three electroactive solutes A, B and C*

If we start at point X on the graph and make E increasingly negative by moving to the right, at E_1 we observe a current as C is reduced.

After E_2 the current is fairly constant (the limiting current) as C is reduced as soon as it reaches the electrode, so that the rate of reduction of C, and therefore the current, is limited by the rate at which C can get to the electrode from other parts of the solution. At E_3 there is a large increase in current as the solvent is reduced. Similar considerations apply to the anodic part of the graph. Oxidation of B (easier than A) begins at E_4 and of A at E_5 (the oxidation current of A is superimposed on the limiting current of B). Finally the solvent is oxidised at E_6.

The principle of ec detection is that we measure the current in a flow cell at the column outlet. We can change the selectivity of the detector by changing the potential of the electrode. For instance, in the figure, if the potential of the electrode was X then the electrode would detect none of the compounds. At point Y the electrode would detect B but not A.

It is difficult in practice to use ec reduction as a method of detection in hplc. Oxygen is very easily reduced, and if it is present in the mobile phase it will create a background current thousands of times larger than the current due to the solutes. To prevent this, oxygen would have to be very carefully removed. This can be done, but it certainly is not easy in practice. So most of the ec applications are oxidations. Another important consideration with ec detectors is that the mobile phase used must have fairly high conductance, so they are used with aqueous-organic mixtures containing added salts, or with buffer solutions.

∏   For which of the following compounds could ec detection be useful?

methylbenzene, decane, phenol, nitrobenzene, 2-chloroaniline

We are looking for compounds that are easy to oxidise, so the hydrocarbons would not be suitable. Nitrobenzene is easily reducible, so ec detection would probably not be useful in practice. Phenols and aromatic amines are easily oxidised, so the last two would be suitable.

One type of ec detector (the *coulometric* detector) reacts all of the electroactive solute passing through it. This type has never become very popular (there is only one on the market at the moment). Another type (the *amperometric* detector) reacts a much smaller quantity of the solute, less than 1%. The currents observed with these detectors are very small (nanoamps), but such currents are not too difficult to measure and the detector has a high sensitivity, considerably higher than that of uv/visible absorbance detectors although not as good as fluorescence detectors. Noise equivalent concentrations of about 10^{-10}g cm^{-3} have been obtained in favourable cases. Another advantage of these detectors is that they can be made with a very small internal volume.

Fig. 2.4j is a simplified diagram of an amperometric detector. Three electrodes are used, called working, auxiliary and reference electrodes (we ae and re). The we is the electrode at which the electroactivity is monitored, and the re, usually a silver–silver chloride electrode, provides a stable and reproducible voltage to which the potential of the we can be referenced. The ae, usually stainless steel, is a current-carrying electrode.

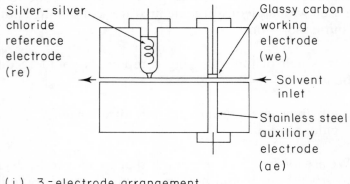

(i) 3-electrode arrangement

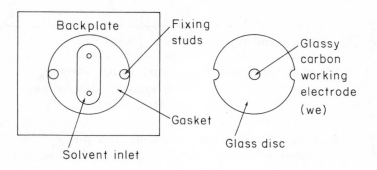

(ii)   Commercial detector

Fig. 2.4j.   *Amperometric detector*

The material most commonly used for the we is glassy carbon, which is a pyrolytically prepared form of carbon that is inert and electrically conducting. The better grades of it have a smooth surface that can accept a high polish. The main problem with solid electrodes like this one is lack of reproducibility caused by degradation of the electrode surface, so that from time to time the electrode surface has to be smoothed, eg by lightly polishing with recorder chart paper or cleaning with chromic acid. Because ec flow cells need this regular maintenance, they have to be easy to get into and reassemble. Fig. 2.4j (*ii*) shows part of the construction of a commercial detector (Waters 460). The glassy carbon we is embedded in a borosilicate glass disc which is clamped on to a backplate. The cell is formed by a ptfe gasket between the disc and the backplate. The cell volume can be varied by changing the thickness of the gasket (the lowest volume for this one is 2.5 μl).

Although they are more sensitive (and cheaper) than uv absorbance detectors, ec detectors are not as easy to use, and have a more limited range of applications. They are chosen for trace analyses where the uv detector does not have a high enough sensitivity. Fig. 2.4k shows some examples of compounds for which ec detection has been used.

Compound type	Examples
phenols, amines	neurotransmitters, eg adrenaline, dopamine; amino-acids
heterocyclic nitrogen compounds	cocaine, morphine alkaloids, phenothiazines, purines
sulphur compounds	penicillins, thioureas, amino acids
unsaturated alcohols	vitamin C
anions	I^-, $S_2O_3^{2-}$, SCN^-

Fig. 2.4k.   *Compounds which can be detected by ec*

SAQ 2.4c

Fig. 2.4l shows the current-potential curves of two electroactive solutes X and Y. In a solution containing both of them:

(*i*)   Which would be detected by an ec detector operating at a potential E_2.

(*ii*) Which would be detected at E_3?

(*iii*) Operating the detector at a potential E_4 would not be a good idea, and operation at E_1 would not be very smart, either. What would be detected at each of these potentials, and what is wrong with the choice of potential in each case?

SAQ 2.4c

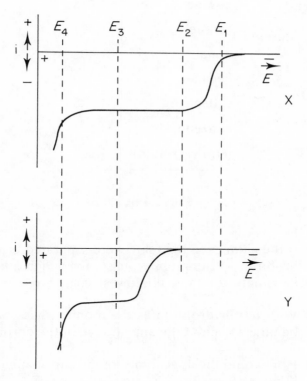

Fig. 2.4l. *Current-potential curves for two electroactive solutes,*
X and Y

2.4.5.   Refractive Index (ri) Detectors

These detectors sense the difference in refractive index between the
column eluent and a reference stream of pure mobile phase. They
are the closest thing in hplc to a universal detector, as any solute
can be detected as long as there is a difference in ri between the
solute and the mobile phase.

∏   Look at the refractive index values given in Fig. 2.4m and
        then see if you can answer the questions below.

hexane	1.375
octane	1.397
nonane	1.405
decane	1.410
tridecane	1.425
benzene	1.501
tetrahydrofuran	1.405

Fig. 2.4m

(*a*) If the alkanes in the table were separated using tetrahy-
       drofuran as the mobile phase, for which alkane would
       an ri detector show the lowest sensitivity?

(*b*) With tetrahydrofuran as the mobile phase, what would
       be unusual about the appearance of the chromatogram?

(*c*) How would the appearance of the chromatogram change
       if benzene was used as the mobile phase?

(*d*) With which mobile phase, benzene or tetrahydrofuran,
       would the ri detector show the greatest sensitivity for
       tridecane?

(*a*) Because nonane has the same ri as that of the mobile phase, the response of the ri detector to nonane would be zero.

(*b*) Hexane and octane have an ri less than that of the mobile phase, whereas for decane and tridecane the ri is greater than that of the mobile phase. The peaks in the chromatogram would appear on either side of the baseline, crossing over at the point where nonane elutes.

(*c*) The ri of benzene is greater than that of any of the alkanes, so that with benzene as the mobile phase the peaks would all be on the same side of the baseline.

(*d*) The difference in ri would be greatest with benzene as the mobile phase, so the use of benzene would give the higher sensitivity (this would apply to all the others as well).

Refractive index detectors are not as sensitive as uv absorbance detectors. The best noise levels obtainable are about 10^{-7} riu (refractive index units), which corresponds to a noise equivalent concentration of about 10^{-6} g cm^{-3} for most solutes. The linear range of most ri detectors is about 10^4. If you want to operate them at their highest sensitivity you have to have very good control of the temperature of the instrument and of the composition of the mobile phase. Because of their sensitivity to mobile phase composition it is very difficult to do gradient elution work, and they are generally held to be unsuitable for this purpose.

Several rather different designs of ri detector have been used in hplc. Fig. 2.4n shows the operating principle of one type, the deflection refractometer. Light from the source S is focussed onto the cell, which consists of sample and reference chambers separated by a diagonal sheet of glass. After passing through the cell, the light is diverted by a beam splitter B to two photocells P1 and P2. A change in the ri of the sample stream causes a change in the relative amounts of light falling on P1 and P2, and therefore a difference in their relative output. This difference is amplified, giving an error signal at the amplifier output that operates a servomotor which rotates the beam splitter until the error signal is reduced to zero. The beam splitter movement (proportional to the difference in ri that caused it) is measured by the recorder.

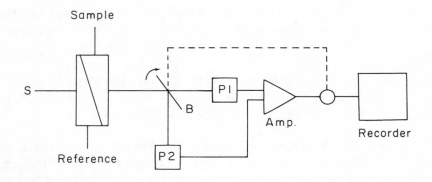

Fig. 2.4n.   *Refractive index detector (refractometer)*

A change in the ri of the sample stream alters the output of P1 and P2, producing a signal at the amplifier output that operates a null-balance system.

2.4.6.   Multipurpose Detectors

Some commercially available detectors have a number of detection modes built into a single unit. Fig. 2.4o is a diagram of the detector used in the Perkin Elmer '3D' system, which combines uv absorption, fluorescence and conductivity detection. The uv function is a fixed wavelength (254 nm) detector, and the fluorescence function can monitor emission above 280 nm, based on excitation at 254 nm. The metal inlet and outlet tubes act as the electrodes in the conductance cell. The detection modes can be operated independently or simultaneously, using a multichannel recorder. In the conductivity mode, using NaCl, a linear range of 10^3 and a noise equivalent concentration of 5×10^{-8} g cm^{-3} have been obtained.

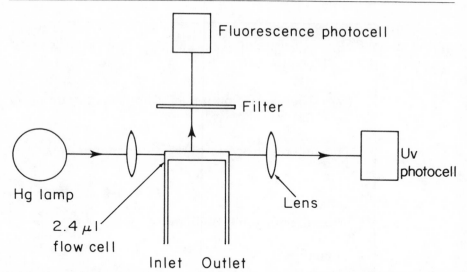

Fig. 2.4o. *Combined uv absorption, fluorescence and conductivity detector. The inlet and outlet tubes are the electrodes in the conductance cell*

SAQ 2.4d

For which of the following analyses do you think that uv absorbance detection would not be suitable? If uv absorbance is unsuitable, suggest an alternative detector.

(*i*)   The determination of mixed sulphonamide drugs in a tablet.

(*ii*) The separation of poly-ethene into fractions of different relative molecular mass, using exclusion chromatography.

(*iii*) The determination of phenols as contaminants in a sample of river water.

$\longrightarrow$

SAQ 2.4d
(cont.)

(*iv*) The analysis of B-vitamins in a multivitamin tablet.

(*v*) The determination of riboflavin (vitamin
B2) in milk.

The general structure of the sulphonamides is:

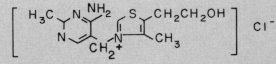

Structures of some B vitamins:

Thiamine(B1)

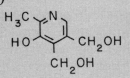

Pyridoxine(B6)

Niacinamide

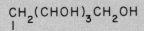

Riboflavin(B2)

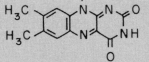

SAQ 2.4d

2.4.7. Derivative Preparation

In gc, derivatives are usually prepared to allow or improve the chromatography of the sample. The purpose of derivative preparation in hplc is usually to improve detection, especially when determining traces of solutes in complex matrices, such as biological fluids or environmental samples.

Derivative preparation can be performed either prior to the separation (pre-column derivatisation) or after (post-column derivatisation), and can be done either on-line or off-line. The two techniques most commonly used are pre-column off-line and post column on-line.

The first of these requires no modification to the instrument and, compared to post-column techniques, has fewer limitations as regards reaction time and conditions. On the other hand, the formation of a reasonably stable and well-defined product is necessary, the presence of excess reagent and by-products may interfere with the separation, and derivatisation may alter the properties of the sample that facilitated separation.

Post-column on-line derivatisation is carried out in a reactor located between the column and the detector. With this technique, the derivatisation reaction does not need to go to completion, provided it can be done reproducibly, and the reaction does not produce any chromatographic interferences. The reaction needs to take place in a fairly short time at moderate temperatures, and the reagent should not be detectable under the same conditions at which the derivative is detected. The mobile phase may not be the best medium in which to carry out the reaction, and the presence of the reactor after the column will increase the extra-column dispersion.

Fig. 2.4p shows three types of post-column reactor. In the open tubular reactor, after the solutes have been separated on the column, reagent is pumped into the column effluent via a suitable mixing tee. The reactor, which may be a coil of stainless steel or ptfe tube, provides the desired holdup time for the reaction. Finally, the combined streams are passed through the detector. This type of reactor is commonly used in cases where the derivatisation reaction is fairly fast. For slower reactions, segmented stream tubular reactors can be used. With this type, gas bubbles are introduced into the stream at fixed time intervals. The object of this is to reduce axial diffusion of solute zones, and thus to reduce extra-column dispersion. For intermediate reactions, packed bed reactors have been used, in which the reactor may be a column packed with small glass beads.

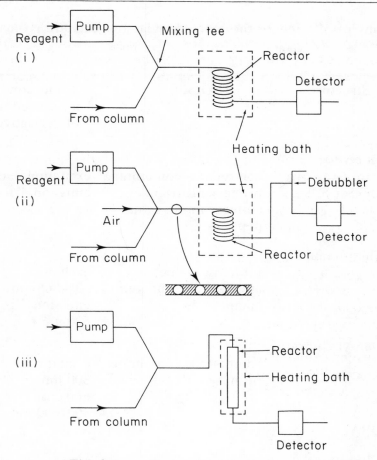

Fig. 2.4p. *Post-column reactors*

(*i*) Open tubular reactor
(*ii*) Segmented reactor
(*iii*) Packed bed reactor

The reagents used in post-column reactors are:

(*a*) Fluorotags: non-fluorescent molecules that react with solutes to form fluorescent derivatives.

(*b*) Chromatags: which form a derivative that strongly absorbs uv or visible radiation.

Examples of a few of the more popular reagents are given in Fig. 2.4q.

Structure	Used for	Conditions for detection of derivative
Fluorotags		
Fluorescamine	compounds containing primary nitrogen, eg amines, amino acids, peptides	excitation 390 nm emission 470 nm
Dansyl chloride	proteins, amines, amino acids, phenolic compounds	excitation 335–365 nm emission 520 nm
OPA	compounds containing primary nitrogen	excitation 300 nm emission 400–600 nm
Chromatags		
Ninhydrin	amino acids	absorption at about 570 nm
PNBDI	carboxylic acids	absorption at about 254 nm

Fig. 2.4q. *Reagents used in post-column reactors*

It is sometimes possible to improve detection by changing the pH of the eluent, or by the use of photochemical reactions. The common barbiturates used in therapy are weak acids that are easily separated in their acid (unionised) forms. Because the conjugate bases are much stronger chromophores than the acids, barbiturates have been detected by post-column mixing with a pH 10 borate buffer followed by uv absorption at 254 nm. An example of the second approach is the detection of cannabis derivatives in body fluids involving the conversion of cannabis alcohols to fluorescent derivatives on subjecting the column effluent to intense uv radiation.

SAQ 2.4e	Consider a solute which is detected by derivatisation, using a post-column reactor of the type shown in Fig. 2.4p(i).
	What would be the effect on the peak area of this solute of:
	(i) Increasing the length of the reactor coil?
	(ii) Increasing the temperature of the reactor coil?
	(iii) Increasing the flow rate of reagent into the mixing tee?
	What would be the effect on the resolution between two peaks in the chromatogram of increasing the length of the reactor coil?

SAQ 2.4e

Summary

A large number of devices have been used as detectors for hplc. The characteristics of five of the more important types are described, and examples are given of the range of samples for which they can be used. The use of derivative preparation as an aid to detection is considered.

Objectives

You should now be able to:

● specify the properties that are required of an hplc detector;

● understand the operating principles and the limitations of the important types of detector;

● recognise the samples for which different detectors can be used;

● describe how derivatives can be prepared to improve detection.

3. Column Packings and Modes of hplc

3.1. INTRODUCTION

3.1.1. The Size and Shape of Column Packings

Fig. 3.1a shows the different sizes and shapes of particles that have been used as stationary phases in hplc. The particles are usually silica, although in ion exchange and exclusion chromatography polymeric gels or resins are common.

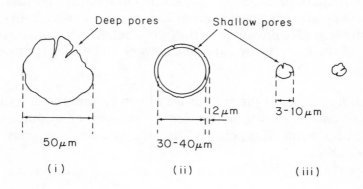

Fig. 3.1a. *Silica particles used in hplc*

(*i*) Large porous particles
(*ii*) Porous layer beads, consisting of a thin shell of silica or modified silica or other material on an inert spherical support
(*iii*) Spherical or irregular microparticulates

The large porous particles are the oldest of these materials, and are no longer used in analytical hplc, although because of their high sample capacity they are still useful in preparative work. Columns packed with the large particles have relatively low efficiencies because of the long time it takes for solute species to diffuse into and out of the porous structure (slow mass transfer).

Porous layer beads (also called pellicular stationary phases) are still used in analytical hplc for some ion exchange applications, and also as guard columns (see Section 5.3.2). The design of the porous layer bead, consisting of an inert solid core of glass or plastic with a thin outer coating of silica, modified silica or some other material, was an early attempt to minimise the time for mass transfer, thus overcoming the problem with large particles. Porous layer beads are more efficient than the large particles but have a smaller surface area and thus a relatively small sample capacity.

Analytical hplc these days is nearly always done with microparticulate column packings, which are small porous particles, usually spherical or irregular silica, with nominal diameters of 3, 5 or 10 μm. They combine the best features of the other two types, having high efficiency as well as a large surface area. In bulk, the appearance of a microparticulate silica resembles that of a fine talcum powder. With microparticulates, dry packing methods result in column beds that are unstable under pressure, so they are packed into columns using a slurry of the material in a suitable solvent and under considerable pressure.

Fig. 3.1b lists some of the properties of two commercially available microparticulate packings. The first is a 5 μm unmodified silica, the second a 3 μm silica, chemically modified with octadecylsilane (ODS, C-18).

	Spherisorb S5W	Spherisorb S3 ODS 2
Type	silica	silica, ODS bonded, fully end-capped
Shape	spherical	spherical
Surface area, $m^2\ g^{-1}$	220	220
Average pore diameter, nm	8	8
Range of pore diameter, nm	5.4–11	5.4–11
Pressure drop, bar	27 (*i*)	212 (*ii*)
Efficiency, plates m^{-1}	60 000–80 000	110 000–150 000
Cost (1986) £/10 g	41.80	71.50

Pressure drops were measured in a 25 cm × 4.6 mm column with a 1 cm^3 min^{-1} mobile flow rate using:

(*i*) hexane/acetonitrile 99 : 1

(*ii*) methanol/water 80 : 20.

Fig. 3.1b. *Properties of two commercially available microparticulate packings*

You will find as you get further involved with hplc that there is a bewildering variety of microparticulate packings available in the trade literature. The textbook by Hamilton and Sewell (second Edition, Chapter 4) has an extensive list, as do most other textbooks. To keep things in perspective, remember that almost all the work in analytical hplc at present is done with chemically modified silicas, ie bonded phases, and of these, by far the most important is the non-polar C-18 type. Even if we restrict ourselves to this one type of packing there is a large selection available, and although they are all designed to do essentially the same job, there are differences between the C-18 packings from different manufacturers, depending on the size, shape and the pore size of the silica particle, the carbon content of the bonded phase, and the extent of end-capping. End-capping is a method used to reduce the residual adsorptive properties of the silica, and is discussed in more detail later.

3.1.2. Modes of hplc

Microparticulate silicas have been used in a number of different ways in hplc (these were mentioned in the introduction):

(*a*) As adsorbents

(*b*) As supports for stationary liquids in partition chromatography

(*c*) As bonded phases

(*d*) As materials for exclusion chromatography.

Methods (*a*) and (*b*) are the original modes and have been replaced to a large extent by the use of bonded phases. Liquid–liquid partition separations, in particular, are very rare these days. Bonded phase chromatography is experimentally much easier than adsorption or liquid–liquid partition. It is more versatile, is quicker, and has better reproducibility than the older modes. In a bonded phase, the highly polar surface of the silica is altered by the chemical attachment of different functional groups. These attached groups can be non polar (eg C-8, phenyl, C-18), polar ($-NH_2$, $-CN$) or ionisable (sulphonic acid, quaternary ammonium). The introduction of ionisable groups produces bonded phases with ion-exchange properties. The range of functional groups that can be bonded to silica is very wide, and for specialised applications (eg the separation of chiral compounds or other isomers) some fairly exotic bonded phases are available.

The terms *normal phase* and *reverse phase* are used to describe adsorption and many bonded phase separations (but not in connection with ion-exchange or exclusion). Normal phase means that the polarity of the stationary phase is higher than that of the mobile phase, which is what happens, for example, when silica is used in adsorption chromatography. Reverse phase means that the polarity of the stationary phase is less than that of the mobile phase, which is the case with hydrocarbon-type bonded phases. Polar bonded phases can be used in either normal or reverse phase modes. With both techniques, solutes are eluted in order of polarity (increasing or decreasing), and we can change the retention times of solutes by changing the polarity of the stationary phase or (more easily) of the mobile phase. These facts are summarised in Fig. 3.1c.

	Normal phase chromatography	Reverse phase chromatography
Stationary phase polarity	High	Low
Mobile phase polarity	Low-medium	Medium-high
Typical mobile phase	heptane/$CHCl_3$	CH_3OH/H_2O
Order of elution	Least polar first	Most polar first
To increase retention of solutes	Decrease mobile phase polarity	Increase mobile phase polarity

Fig. 3.1c. *Summary of characteristics of normal and reverse phase chromatography*

For ionisable solutes, the pH of the mobile phase is an important factor in the control of retention and selectivity.

Fig. 3.1d shows how an hplc method is chosen on the basis of the solubility of the sample and the sort of functional groups that it contains. The figure deals only with silica or modified silica stationary phases; remember that for some modes (ion-exchange or exclusion) stationary phases are available that are based on materials other than silica. Samples having components with relative molecular masses greater than about 2000 would be analysed by exclusion chromatography. You will see from the Fig. that there is always a choice of modes and also that almost any separation can be achieved by reverse phase chromatography using a bonded silica stationary phase. This is the mode that we would tend to look at first; it is often faster, cheaper and experimentally easier than the alternatives.

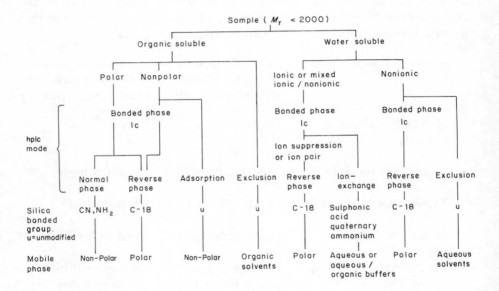

Fig. 3.1d.   *Choice of an hplc mode*

Fig. 3.1e is a fairly recent summary of the amount of use of the different modes. It was obtained by surveying 369 papers on hplc applications from 10 different journals. You can see that bonded phases were used in nearly 80% of the applications, and that C-18 bonded phases were used in over half of them. Since these data were collected the use of C-18 columns has increased; they are probably now used in around 75% of applications, as suggested in the introduction.

	% column usage, 1982–3
Bonded phases	78
reverse phase, C-18	(54)
reverse phase, others	(18)
normal phase	(6)
Adsorption	10
Ion-exchange	6
Exclusion	6

Fig. 3.1e.   *Use of different modes of hplc. Source: Analytical Chemistry Reviews, April 1984*

3.1.3.   The Importance of Polarity in hplc

The relative distribution of a solute between two phases is determined by the interactions of the solute species with each phase. The relative strengths of these interactions are determined by the variety and the strengths of the intermolecular and other forces that are present, or, in more general terms by the polarity of the sample and that of the mobile and stationary phases.

Intermolecular forces may be caused by a solute molecule having a dipole moment, when it can interact selectively with other dipoles. If a molecule is a good proton donor or acceptor it can interact with other such molecules by hydrogen bonding. Molecules can also interact via much weaker dispersion forces which rely on a given molecule being polarised by another molecule.

Polarity is a term that is used in chromatography as an index of the ability of compounds to interact with one another in these various ways. It is applied very freely to solutes, stationary phases and mobile phases. The more polar a molecule, the more strongly it can interact with other molecules through the mechanisms above. If the polarities of stationary and mobile phases are similar, then it is likely that the interactions of solutes with each phase may also be similar, resulting in poor separations. Thus, for hydrocarbon-type (non-polar) stationary phases, we use a polar mobile phase, whereas unmodified silica, which is highly polar, needs a mobile phase with relatively low polarity. If we are concerned with the separation of solutes that are chemically very similar, we would try to choose a stationary phase that is chemically similar to our solutes. Retention of solutes is usually altered by changing the polarity of the mobile phase.

It is easy to see that, for instance, water is a more polar solvent than heptane. Water has a dipole moment, is both a proton donor and acceptor and will dissolve ionic solutes. Similarly, methanol and acetonitrile are both more polar than heptane, but it is not so easy to assign relative polarities to methanol and acetonitrile.

It is helpful in lc to have a quantitative measure of polarity, so that, for example, the relative polarity of a solvent or a mixture of solvents can be expressed as a number. There are several ways in which this has been done; none of them are entirely satisfactory, but they do allow us to arrange solvents in order of polarity and to make rough estimates of the polarity of solvent mixtures. One such way is to use as a measure of polarity a quantity called the solubility parameter, δ, defined by:

$$\delta = (\Delta E / V)^{\frac{1}{2}} \qquad (3.1a)$$

where ΔE = internal energy of vaporisation and V = molar volume.

In SI units, δ is measured in $J^{\frac{1}{2}} m^{-\frac{3}{2}}$, or $Pa^{\frac{1}{2}}$, although many authors still give δ in the non-SI $cal^{\frac{1}{2}} cm^{-\frac{3}{2}}$ (in the old units, δ values range from about 7, for hydrocarbons, to about 23, for water). For polar solvents, there are strong intermolecular forces in the liquid state leading to large values of ΔE and δ.

Another measure of polarity, the polarity index, P', is calculated from solubility data. This quantity again increases with increasing polarity. Fig. 3.1f shows the values of both of these quantities for a range of solvents, in order of increasing P'. The order of polarity is not quite the same by the two methods. Although it is not strictly correct, we can make rough estimates of the polarity of mixtures on the basis that the polarity of the mixture varies linearly with composition.

Solvent	Solubility parameter δ, $Pa^{\frac{1}{2}}$	Polarity index P'
hexane	14.9	0
methylbenzene	18.2	2.3
dichloromethane	19.8	3.4
1,2-dichloroethane	20.0	3.7
tetrahydrofuran	18.6	4.2
ethyl ethanoate	19.6	4.3
trichloromethane	19.0	4.4
butanone	19.0	4.5
dioxane	20.4	4.8
ethanol	25.9	5.2
propanone	20.2	5.4
acetonitrile	23.9	6.2
2-methoxyethanol	23.3	6.3
methanol	29.4	6.6
water	47.8	9.0

Fig. 3.1f. *Solubility parameter and polarity index for a range of solvents*

∏ What would be the P' value of a 50:50 methanol/water mixture?

$$P' = (6.6 \times 0.5) + (9 \times 0.5) = 7.8$$

If a particular combination of stationary and mobile phases produces satisfactory retention times but the solutes are not completely separated, we may be able to improve the selectivity without too much change in the retention times by formulating a solvent mixture with the same polarity but using different solvents. What mixture of tetrahydrofuran and water would have the same value of P' as the methanol/water mixture above?

If $v = $ volume fraction H_2O,

$$7.8 = 9 \times v + 4.2(1 - v).$$

Hence, the proportions of tetrahydrofuran and water would be 25% tetrahydrofuran, 75% water.

Changing the organic component of the mobile phase can sometimes have a profound effect on selectivity. Fig. 3.1g shows a case where the order of elution is reversed when the mobile phase is changed from methanol/water 50:50 to tetrahydrofuran/water 25:75.

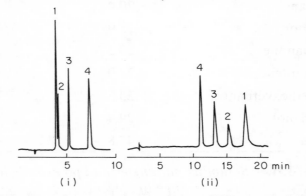
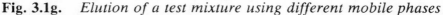

Fig. 3.1g. *Elution of a test mixture using different mobile phases*

Column:	5 μm C-8 bonded phase 15 cm × 4.6 mm
Mobile phase:	(*i*) methanol/water 50:50
	(*ii*) tetrahydrofuran/water 25:75
Flow rate:	1 cm^3 min^{-1}
Detector:	Uv absorption, 254 nm
Sample:	1 = 4-nitrophenol 2 = nitrobenzene
	3 = 1,4-dinitrobenzene
	4 = methyl benzoate

It is possible to alter selectivity by the use in the mobile phase of more than one organic component, by adjustment of pH or ionic strength, or by the use of ion-pair reagents or other complexing agents. Mobile phases containing more than one organic component are composed of solvents that can interact to different extents with different components of the sample. Selectivity can be optimised by varying the mobile phase composition (ie varying the degree of interaction between individual sample components and the mobile phase). Such development work is conveniently done using 3 × 3 columns (Section 2.3.7) which equilibrate rapidly following a change in mobile phase composition.

Buffer solutions are used to control retention and selectivity in the chromatography of ionisable solutes, and in addition the chemical nature of the buffering agent can affect secondary equilibria, eg interaction of the solute with silanol groups (see Section 4.2).

Summary

Microparticulate silica can be used in a number of modes for hplc; of these, reverse phase chromatography using bonded phases is the most widely used. In normal and reverse phase chromatography the retention times and selectivities of solutes can be altered by adjustment of the nature and composition of the mobile phase.

Objectives

You should now be able to:

● specify the size and shape of column packings used in hplc;

● describe the modes in which these packings can be used;

● identify a mobile phase suitable for each mode;

● distinguish between normal and reverse phase chromatography, and understand how the retention times of solutes are influenced by mobile phase polarity in these techniques;

- arrange solvents in approximate order of polarity, and estimate the polarity of mixtures.

3.2. BONDED PHASE CHROMATOGRAPHY

3.2.1. The Nature and Preparation of Bonded Phases

The method most commonly used to prepare bonded phases from silica involves reaction of the silica with a substituted dimethylchlorosilane. Fig. 3.2a shows the reaction, in which HCl is eliminated between a surface silanol group and the silylating agent.

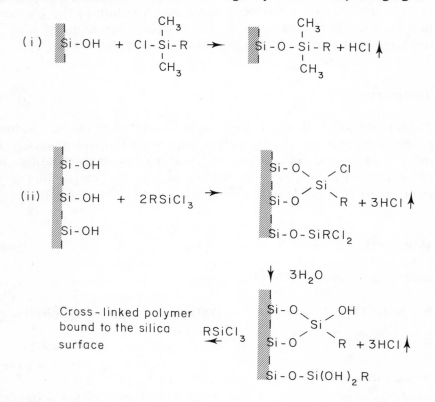

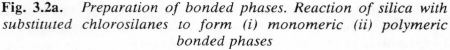

Fig. 3.2a. *Preparation of bonded phases. Reaction of silica with substituted chlorosilanes to form (i) monomeric (ii) polymeric bonded phases*

Before reaction, the silica is treated with acid (eg refluxed for a few hours with 0.1 mol dm^{-3} HCl). This treatment produces a high concentration of reactive silanol groups at the silica surface, and also removes metal contamination and fines from the pores of the material. After drying, the silica is then refluxed with the dimethylchlorosilane in a suitable solvent, washed free of unreacted silane and dried. This reaction produces what is called a 'monomeric' bonded phase, as each molecule of the silylating agent can react with only one silanol group.

More complicated surface structures can be produced by changing the functionality of the silylating agent and the conditions under which the reaction is carried out. The use of di- or trichlorosilanes in the presence of moisture can produce a crosslinked polymeric layer at the silica surface, as shown in Fig. 3.2a (*ii*). Monomeric bonded phases are preferred, as their structure is better defined and they are easier to manufacture reproducibly than the polymeric materials.

The different kinds of bonded phase can be made by varying the nature of the functional group R in the silylating agent. For example, a bonded phase cation-exchanger can be made by using for R a phenyl or phenyl substituted alkyl group. After the bonding reaction, the phenyl is sulphonated using chlorosulphonic acid. An anion-exchanger can be made by using for R a chlorinated alkyl group, which then forms a quaternary salt by reaction with a tertiary amine.

Many other methods have been used to prepare bonded phases; these include esterification of the surface silanol groups with alcohols, or conversion of the silanol groups to Si—Cl using thionyl chloride, followed by reaction with an organometallic compound. If you are interested, there are details in the textbooks by Knox or by Hamilton and Sewell.

It is not possible to bond all of the surface silanol groups. Unreacted silanols are capable of adsorbing polar molecules, and will thus affect the chromatographic properties of the bonded phase. Usually, the unreacted silanols produce undesirable effects, such as tailing and excessive retention in reverse phase separations, although there have been cases reported where the unreacted silanols improve such separations. The concentration of unreacted silanols in non-polar bonded phases is normally reduced by the process of 'end-capping', in which most of the remaining silanols are reacted with a small silylating agent, such as trimethylchlorosilane.

Fig. 3.2b shows the surface structure of an ODS bonded phase, containing bonded C-18 alkyl groups, end-capped silanol groups and a small number of free silanols.

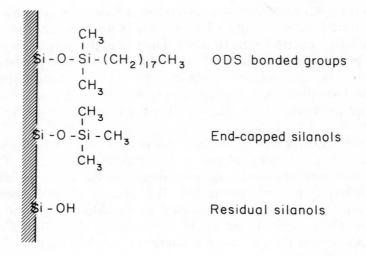

Fig. 3.2b.   *Structure of an ODS (C-18) surface*

Fig. 3.2c shows a selection of the commercially available bonded phases. There are comprehensive lists of these in the two textbooks mentioned earlier.

Name	Attached functional group	Type	Particle shape and size, μm	Price £/10 g (1986)	Supplier
Hypersil SAS	C-1	NP	spherical, 5	80	Shandon Southern
Spherisorb S5P	Phenyl	NP	spherical, 5	64.90	Phase Separations
Lichrosorb RP-8	C-8	NP	irregular, 10	72.0	E. Merck
Nova-Pak C-18	C-18	NP	spherical, 4	(*i*)	Waters
Nucleosil C-18	C-18	NP	spherical, 5 or 10	75.80 (both)	Macherey Nagel
Partisil 10 SAX	quaternary ammonium	SAX	irregular, 10	107	Whatman
Partisil 10 SCX	sulphonic acid	SCX	irregular, 10	107	Whatman
Lichrosorb CN	$-(CH_2)_3CN$	WP	irregular, 10	74	E. Merck
Spherisorb S3NH$_2$	$-(CH_2)_3NH_2$	P or WAX	spherical, 3		Phase Separations

(*i*) available only as packed columns

NP, WP, P = non-polar, weakly polar, polar, respectively
SCX, SAX, WAX = strong cation, strong anion, weak anion exchanger, respectively.

Fig. 3.2c. *Commercially available bonded phases*

3.2.2. Normal and Reverse Phase Chromatography

Reverse phase operation with bonded phases has achieved very wide
popularity because it has the following advantages:

(*a*) The mode has a very broad scope that allows samples with wide
ranges of polarity to be separated,

(*b*) It can be applied to the separation of ionic or ionisable compounds by the use of ion pairing or ion suppression techniques
(see Section 3.3.3),

(*c*) The mode is generally experimentally easier, faster and more
reproducible than other 1c modes.

The mode is, however, not without limitations. Among the more
important of these are:

(*a*) For silica bonded phases, stable columns can be maintained
only over a pH range between about 3 and 8. Below pH 3 the
bonded group may be removed, and above pH 8 the silica is
appreciably soluble in the mobile phase. It is possible to operate
at high pH, at least for short periods of time, by presaturating
the mobile phase with silica. The limited pH range is not a
serious handicap, as most separations can be carried out within
these limits.

(*b*) The presence of unreacted silanol groups on the silica surface can often cause poor peak shape and non reproducible
behaviour between columns, due to solute adsorption. These
effects can in many cases be overcome, as will be seen later.

(*c*) The reverse phase retention mechanism is still not properly
understood. One possible explanation is that the hydrophobic
surface of the bonded phase extracts the less polar constituent
of the mobile phase to form a layer at the silica surface, and
that partitioning of solutes then occurs between this layer and
the mobile phase. However, in many cases in reverse phase
chromatography there may be several mechanisms operating
at the same time (adsorption on unreacted silanol groups, for

instance). On the basis that a better understanding of separation mechanisms with bonded phases will expand their uses, there is much study currently directed towards this problem.

Solute retention in reverse phase separations is governed by a number of effects, the more important of which were listed in Fig. 3.1c. Solutes are eluted in order of polarity, the most polar being eluted first. If we think of a reverse phase separation as being a partitioning process in which solutes are distributed between a non-polar stationary phase and a polar mobile phase then the non-polar solutes will be soluble in the stationary phase, and will travel through the system more slowly than the polar solutes, which favour the mobile phase. We can change the distribution by changing the polarity of the mobile phase. For instance, if we make the mobile phase less polar (by increasing the ratio of organic solvent to water) then we will shift the distribution of the solutes towards the mobile phase, and their retention will decrease. We can also alter retention by changing the polarity of the stationary phase, for instance by using bonded phases with different nonpolar groups.

∏   What would be the effect on the retention of a given solute of:

   (*a*) Changing from a C-18 to a phenyl bonded phase?

   (*b*) Changing from a C-18 bonded phase to another C-18 that contained a smaller percentage of carbon?

   (*a*) The phenyl bonded phase is slightly more polar than the C-18, so we would expect retention to decrease,

   (*b*) Similarly, decreasing the percentage of carbon in a given type of bonded phase will increase the polarity, so that, other things being equal, we would expect retention to decrease here as well. In commercial C-18 packings the carbon content of the bonded phase can be between 5 and 20%.

Fig. 3.2d is a chromatogram of a test mixture on (*i*) a C-18 bonded phase and (*ii*) a —CN bonded phase. The —CN is a weakly polar bonded phase that can be used either in normal or reverse phase mode. In each case the solute with the highest polarity (the alcohol) is eluted first and the solute with the lowest polarity (the ether) eluted last. Increasing the polarity of the bonded phase on going from C-18 to —CN causes the retention of all solutes to decrease.

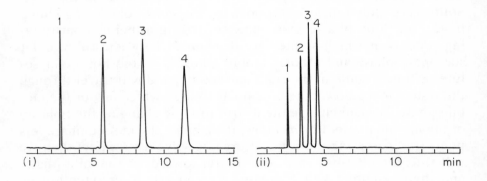

Fig. 3.2d.   *Reverse phase chromatogram of test mixture*

Column:	(*i*) 4 μm C-18
	(*ii*) 4 μm-CN, both 15 cm × 3.9 mm
Mobile phase:	CH_3CN/H_2O 40 : 60
Flow rate:	2 cm^3 min^{-1}
Sample:	1 = benzyl alcohol
	2 = 2-phenoxyethanol
	3 = 4-methoxybenzaldehyde
	4 + methyl phenyl ether
Detector:	Uv absorption, 254 nm.

Fig. 3.2e shows a reverse phase separation of some tricyclic antidepressant drugs. These compounds are weak bases, and at the pH used will be completely protonated (see Section 3.3.1). Because the protonated bases are very polar compounds, they are adsorbed strongly by unreacted silanol groups, causing excessive retention and severe peak tailing. Effects like this are possible even with end-capped bonded phases. There are several ways to approach the problem. One is to add to the mobile phase a large concentration (relative to the concentration of the solutes) of a competing base. Because of its relatively high concentration it is preferentially

adsorbed by the silanol groups thus minimising the adsorption of the other bases. The figure shows the dramatic change that results when nonylamine is added as a competing base. Other possibilities for this separation would be to use ion suppression or ion pairing techniques (Section 3.3.3).

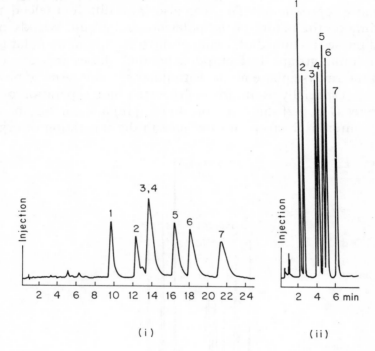

Fig. 3.2e. *Separation of antidepressant drugs*

Column:	C-8 bonded phase 15 cm × 4.6 mm
Mobile phase:	(*i*) $CH_3CN/0.01$ mol dm^{-3} H_3PO_4 adjusted to pH 2.5 with KOH
	(*ii*) $CH_3CN/0.01$ mol dm^{-3} H_3PO_4 + 0.005 mol dm^{-3} nonylamine, pH 2.5
Flow rate:	2 cm^3 min^{-1}
Detector:	Uv absorption, 254 nm
Sample:	1 = nordoxepin 2 = doxepin 3 = desipramine 4 = protriptylene 5 = imipramine 6 = nortryptyline 7 = amitriptyline

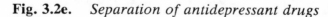

Operation of bonded phases in the normal phase mode is now
replacing many separations that were previously done by adsorption chromatography with unmodified silica. Compared to silica,
the bonded phases show less tailing and respond more rapidly to
changes in mobile phase composition. The chromatography is generally more reproducible. They can also show different selectivities,
depending on the nature of the polar bonded group. Weakly polar
bonded phases include diol, cyano or nitro groups; more polar types
contain amino groups. In fact, polar bonded phases are most often
used in the reverse phase mode, with polar solvents. Amine bonded
phases are especially useful for the reverse phase separation of carbohydrates (Fig. 3.2f shows a typical example) and can also be used
as weak anion exchangers, for instance in the separation of organic
acids.

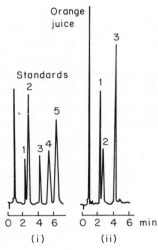
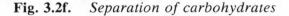

Fig. 3.2f.   *Separation of carbohydrates*

Column:	$5\mu m$ $-NH_2$ bonded phase, 25 cm × 4.6 mm
Mobile phase:	CH_3CN/H_2O 78 : 22
Flow rate:	2 cm^3 min^{-1}
Detector:	refractive index
Chromatograms:	(*i*) standards
	(*ii*) orange juice diluted in two parts of CH_3CN
Peaks:	1 = fructose 2 = glucose 3 = sucrose 4 = maltose
	5 = lactose

3.2.3. Chiral Stationary Phases

The separation of chiral compounds by hplc is a technique that is becoming increasingly important. Chiral compounds can be used in the synthesis of biologically active materials, thus there is a need for methods to determine optical purity. In some cases, enantiomers of pharmaceutical compounds can differ substantially in their physiological activity. For example, the dextrorotatory enantiomer (S)-(+)-methamphetamine (N-methyl-1-phenyl-2-aminopropane) is a controlled substance with a considerable history of drug abuse, whereas the (R)-(−)-enantiomer is far less potent, and is an ingredient of proprietary decongestants. In the pharmaceutical industry, regulatory authorities may require information on the enantiomeric composition of drugs.

A number of specialised stationary phases have been developed for the separation of chiral compounds. They are known as chiral stationary phases (CSPs) and consist of chiral molecules, usually bonded to microparticulate silica. The mechanism by which such CSPs discriminate between enantiomers (their *chiral recognition*, or *enantioselectivity*) is a matter of some debate, but it is known that a number of competing interactions can be involved. Columns packed with CSPs have recently become available commercially. They are some three to five times more expensive than conventional hplc columns, and some types can be used only with a restricted range of mobile phases. Some examples of CSPs are given below:

(*a*) 'Pirkle' CSPs (named after W.H. Pirkle, University of Illinois, who pioneered their development). These use chiral phenylglycine or leucine derivatives, bonded to 5 μm silica, and are available in ionic or covalent forms, the structures of which are shown in Fig. 3.2g (*i*). Generally, the ionic CSP will retain a given chiral sample more than the corresponding covalent form, but the ionic types can be used only with nonaqueous mobile phases of fairly low polarity (hydrocarbons modified with small amounts of alcohols). The covalent CSPs have no such restriction.

(*b*) Some proteins are known to interact selectively with chiral compounds; another type of CSP uses bovine serum albumin bonded to microparticulate silica. A phosphate buffer modified with small amounts of 1-propanol is used as the mobile phase.

(*c*) CSPs based on ligand exchange. These use a bonded chiral amino acid complexed with Cu^{2+}, and a copper salt (about 10^{-3} mol dm^{-3}) modified with organic solvents as the mobile phase. They have been used to separate α-amino acids and other chiral compounds which can form chelate complexes with Cu^{2+}.

(*d*) Cyclodextrin bonded phases. Cyclodextrins are cyclic chiral carbohydrates containing from six to twelve glucose units. The CSPs use α, β or γ cyclodextrin (having 6, 7 or 8 glucose units, respectively), bonded to 5 μm silica. The shape of a cyclodextrin molecule resembles an open barrel, with an internal cavity whose dimensions are determined by the number of glucose units in the molecule. Solutes are separated by a mechanism that involves the formation of inclusion complexes, the strengths of which are governed mainly by the ability of the solute to fit into the cyclodextrin cavity. Thus the stationary phase can discriminate between solutes that differ only in geometry or spatial orientation, and has found use in the separation of a number of structural and positional isomers, as well as enantiomers. A simplified diagram of the inclusion process is shown in Fig. 3.2g (*ii*).

The mobile phases used with cyclodextrin CSPs are similar to those used in reverse phase chromatography, ie water or buffer solutions modified with methanol or acetonitrile.

∏   The organic phase modifier competes with the solute molecules for inclusion in the cyclodextrin cavity. How would an increase in the concentration of modifier affect the retention time of a solute?

Increasing the concentration of organic modifier will decrease the interaction between the solute and the cavity and will thus lead to a shorter retention time.

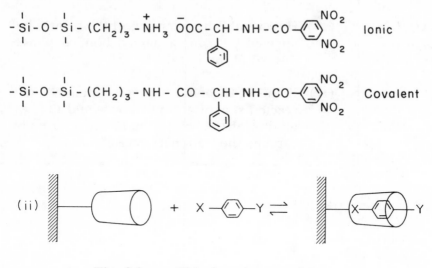

Fig. 3.2g. *Chiral stationary phases*

(*i*) Structure of Pirkle CSPs (3,5-dinitrobenzoylphenylglycine bonded silicas)

(*ii*) Inclusion of a solute on a cyclodextrin bonded phase. The interior cavity of the cyclodextrin is relatively non-polar so that the less polar of the substituent groups (X) is oriented into the cavity.

SAQ 3.2a

A test mixture consisting of phenyl methyl ketone, nitrobenzene, benzene and methylbenzene is to be separated on a C-18 column with a mobile phase of CH_3OH/H_2O 60:40. With these conditions, the ketone is eluted first.

(*i*)   In what order are the other solutes eluted?

(*ii*) How would you change the composition of the mobile phase so as to increase the retention of the solutes? $\longrightarrow$

SAQ 3.2a
(cont.)

(*iii*) How would the retention of the solutes be affected by using a phenyl bonded phase instead of the C-18?

(*iv*) If the C-18 bonded phase contained un-reacted silanol groups, how would the retention of the solutes be affected by end-capping the stationary phase?

SAQ 3.2a

Summary

The surface of microparticulate silica can be modified by the attachment of different groups to produce bonded phases. Reverse phase chromatography using bonded phases is generally faster and easier than other modes, and consequently has achieved very wide popularity.

Objectives

You should now be able to:

- Describe the methods used for the preparation of bonded phases;

- predict how solute retention times on bonded phases will be affected by a change in the polarity of the stationary or mobile phase;

- appreciate that chiral solutes can be separated on suitable bonded phases.

3.3. THE CHROMATOGRAPHY OF IONIC SOLUTES

3.3.1. Ion-exchange Chromatography

An ion-exchange resin consists of an insoluble, rigid 3-dimensional matrix, the surface of which has ionisable sites which can carry a positive or a negative charge. Each site also requires an oppositely charged ion (the counter ion) to preserve overall charge neutrality. If these ionisable sites are positively charged, the counter ion is an anion and the resin will attract and exchange anions from solution:

$$R^+Y^- + X^-(\text{solution}) \rightleftharpoons R^+X^-(\text{resin}) + Y^-(\text{solution})$$

In this scheme, a counter ion Y^-, attached to the R^+ group on the resin, is exchanged for another anion X^- from solution. To exchange cations, we need a structure with exchange sites that are negatively charged, and associated positive counter ions:

$$R^-Y^+(\text{resin}) + X^+(\text{solution}) \rightleftharpoons R^-X^+(\text{resin}) + Y^+(\text{solution})$$

Ion-exchange chromatographic separations are based on different strengths of interaction between the solute ions in the mobile phase and the fixed exchanging groups on the stationary phase.

The terms *strong* or *weak* as applied to ion-exchange resins indicate how the exchanging properties of the structure vary with pH. Strong ion-exchangers contain strongly acid or basic groups that are fully ionised over a wide range of pH. A strong cation-exchange (scx) resin contains sulphonic acid exchange sites that are fully ionised above about pH 2. A strong anion-exchange (sax) resin contains quaternary ammonium exchange sites that are fully ionised up to about pH 10. Weak cation and anion-exchangers contain, respectively, carboxylic acid and amino groups. These materials have a higher exchange capacity then the strong types, but they are ionised only over a restricted range of pH. This behaviour is shown in Fig. 3.3a.

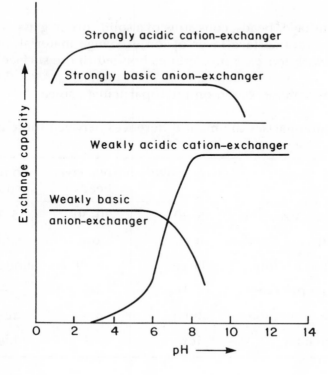

Fig. 3.3a. *Variation of ion-exchange capacity with pH*

The 'weak' exchangers have a higher capacity than the 'strong' exchangers when fully ionised, but are only fully ionised over a restricted pH range.

The capacity of an ion-exchanger measures the amount of material that can be exchanged by a given amount of the resin, and is normally expressed in milliequivalents per g of dry resin.

Classical ion-exchange resins are styrene-divinylbenzene copolymers to which ionisable functional groups are attached. There are several ion-exchange materials that are used for hplc:

(*a*) Microparticulate resins based on styrene-divinylbenzene co-polymers

(*b*) Porous layer beads, consisting of a solid core of glass or polymer with a thin surface layer of ion-exchange material, or a silica bead with ion-exchange groups bonded to the surface

(*c*) Bonded phases based on microparticulate silica.

Fig. 3.3b summarises the main differences between these materials.

	styrene-dvb	porous layer beads	bonded silica
Particle size, μm	5–20	30–50	5–10
Capacity	high	low	high
Sample loading	large	small	moderate
Usable pH range	2–14	2–9	2–8
Packing method	slurry	dry	slurry
Efficiency	low	$\rightarrow$	high

Fig. 3.3b. *Ion-exchange materials used for hplc*

The styrene-dvb types have better temperature stability than the other two, but most resins swell in contact with solvents.

∏ Weak cation-exchangers are not available based on silica. Why not? (refer to Fig. 3.3a).

They are only fully ionised at high pH, when silica is appreciably soluble.

The textbook by Hamilton and Sewell has a list of commercially available ion-exchange materials for hplc.

Retention of solutes in ion-exchange chromatography is determined by the nature of the sample, the type and concentration of other ions present in the mobile phase, the pH, temperature, and the presence of solvents. Because there are so many variables, it is often not easy to predict what will happen in an ion-exchange separation if we change the experimental conditions. There are some useful guidelines, and to see how they work we will look at the ion-exchange separation of two weak acids (see Fig. 3.3c).

Benzoic acid

2-methylbenzoic acid

$$C_6H_5 COOH \rightleftharpoons C_6H_5COO^- + H^+$$

$$K_a = \frac{[H^+][C_6H_5COO^-]}{[C_6H_5COOH]} = 6.3 \times 10^{-5} \qquad K_a = 2 \times 10^{-4}$$

$$pK_a = -\log K_a = 4.2 \qquad\qquad pK_a = 3.7$$

Fig. 3.3c. *Weak acids* it

∏   The acids are both partly ionised in aqueous solution. What would be the effect on the dissociation of the acids if they were made up in (*a*) an acid buffer (*b*) an alkaline buffer?

Adding a common ion (eg H^+) shifts the equilibrium to the left, so if the pH of the buffer is low enough the weak acids will be present as neutral molecules. Similarly, we can force the acids to ionise completely if the pH is high enough. In between, both forms will be present.

We want the acids to be present only as ions, as the two forms of each acid will have different retentions on the ion-exchange stationary phase. As a rough guide, to suppress ionisation completely, we want to buffer at pH = (pK_a − 1.5) and to cause complete ionisation we need pH = (pK_a + 1.5).

For weak bases, the rule operates the other way round. To suppress ionisation we use $pH = (pK_b + 1.5)$ and to ionise completely we need $pH = (pK_b - 1.5)$.

In a pH 5.7 ethanoate buffer, the two weak acids will be fully ionised. In this buffer, separation will occur because of competition for the stationary phase exchange sites between the ethanoate ions and the acid anions:

$$R^{+-}OOCCH_3 + {}^-OOC \qquad \rightleftharpoons R^{+-}OOC \qquad + CH_3COO^-$$

(resin)   (solution)   (resin)   (solution)

∏   If the pH was kept constant but the ethanoate concentration in the mobile phase was increased, would the retention of the weak acids increase, decrease or remain the same?

Increasing the concentration of ethanoate in the mobile phase will force the equilibrium above to the left. The concentration of benzoate in the mobile phase will increase, so the retention will decrease. Similarly, we could increase the retention by using a lower concentration of ions in the mobile phase.

If the pH of the mobile phase is changed, then we will alter the proportions of the ionised and unionised forms of each acid. Decreasing the pH will increase the amount of the unionised form of each acid and so reduce the retention. The stronger acid (with the smaller pK_a) will be the more ionised and so will be retained longer by the stationary phase.

The behaviour will also be affected by the type of salt used to form the buffer. Generally, the ions most strongly retained by the resin are those with a large charge and a small size.

∏   If the ethanoate in the buffer was replaced by citrate, and there were no other competing equilibria, what would be the effect on the retention of the two acids?

The citrate would be strongly retained by the stationary phase, so that the retention of the two acids would decrease. However, if mobile phase buffers are formed using polyvalent salts, there is a strong probability of complex formation, which will alter the predicted behaviour of the system.

Ion exchange equilibria are usually established or altered much faster at higher temperatures. Increasing the temperature will improve efficiency, decrease retention and may alter the selectivity of the separation. The use of organic solvents in the mobile phase will also cause retention to decrease, but because the use of solvents will change many of the variables in an ion-exchange separation, their other effects are not easy to predict.

Fig. 3.3d(*i*) shows the chromatogram of the two acids on a weak anion-exchange column. At the pH used, there will be appreciable amounts of each acid in the unionised form, which probably accounts for the poor shape of the peaks. Addition of an organic solvent to the mobile phase, (*ii*), greatly decreases the retention and also alters the selectivity.

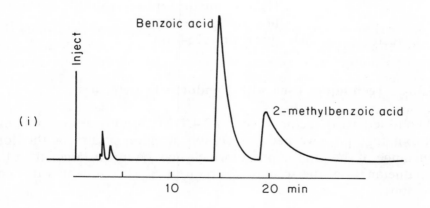

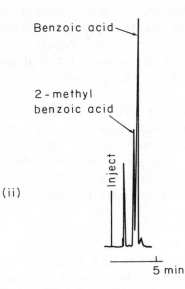

Fig. 3.3d. *Ion-exchange separation of weak acids*

Column:	10 μm —NH_2 bonded phase, 30 cm × 4 mm (weak anion exchanger).
Mobile phase:	(*i*) 0.01 mol dm^{-3} KH_2PO_4, pH 4.2
	(*ii*) 0.01 mol dm^{-3} KH_2PO_4 in H_2O/tetrahydrofuran 50 : 50 flow rate 2 cm^3 min^{-1}
Detector:	Uv absorption, 254 nm.

3.3.2. Techniques Used with Conductivity Detectors

If conductivity detectors (Section 2.4.4) are used with ion-exchange chromatography, we need to be able to compensate for the conductance of the mobile phase, which may be much larger than the conductances of the solutes. There are two ways in which this can be done:

(*a*) Suppressor columns. Consider the separation of cations, eg alkali metal cations on an ion-exchange column using dilute HCl as the mobile phase. The separated ions elute from the analytical column in a background of dilute HCl. They then pass through the suppressor column, which is a strong anion-exchange resin in the OH$^-$ form. This converts the alkali metal chlorides to hydroxides and the HCl mobile phase to H_2O. The reactions are:

Analytical column:

$$\text{Resin-SO}_3^- \text{H}^+ + \text{M}^+ \rightleftharpoons \text{Resin-SO}_3^- \text{M}^+ + \text{H}^+$$

Suppressor column:

$$\text{Resin-NR}_3^+ \text{OH}^- + \text{M}^+\text{Cl}^- \rightleftharpoons \text{Resin-NR}_3^+ \text{Cl}^- + \text{M}^+\text{OH}^-$$

$$\text{Resin-NR}_3^+ \text{OH}^- + \text{H}^+\text{Cl}^- \rightleftharpoons \text{Resin-NR}_3^+ \text{Cl}^- + \text{H}_2\text{O}$$

The action of the suppressor column is thus to reduce the conductance of the mobile phase to a low level, by removing the HCl.

∏   Can you think of the corresponding arrangement that could be used for anions?

We would need an anion-exchange resin in the analytical column, a cation-exchanger in the suppressor column, and a mobile phase containing NaOH. In the suppressor column, Na$^+$ would be exchanged for H$^+$.

In practice, other mobile phases are preferred to NaOH or HCl, for instance amine hydrochlorides (for cations) or $Na_2CO_3/NaHCO_3$ (for anions).

∏   There are two fairly obvious disadvantages associated with the use of suppressor columns; can you see what they are?

The disadvantages are that the suppressor column will increase the dispersion of the system, and it will also have to be regenerated from time to time.

(*b*) Unsuppressed ion-chromatography. In this technique a stationary phase with a very low exchange capacity is used, together with a mobile phase containing ions that interact strongly with the stationary phase. This allows the use of a dilute mobile phase with a low conductance, eg potassium hydrogen phthalate 5×10^{-4} mol dm^{-3} (conductivity about 75 μS cm^{-1} at room temperature). Unsuppressed ion-chromatography, however, has lower sensitivity and usually a smaller linear range than methods using suppressor columns. At present, both techniques are used.

3.3.3.   Ion-suppression and Ion-pair Chromatography

Many separations previously done by ion-exchange are now achieved more easily by the use of ion-suppression or ion-pairing techniques. Ion-suppression is used for the chromatography of weak acids or bases. The principle is that we suppress the ionisation of an acid or the protonation of a base by adjusting the pH, and then chromatograph the sample on a reverse phase column (eg C-18) using methanol or acetonitrile plus a buffer solution as the mobile phase. The technique is preferable to ion-exchange because, compared to ion-exchange columns the C-18 column has higher efficiency, equilibrates faster and is generally better behaved.

Ion-pairing techniques are also used to separate weak acids and bases but additionally they find application in the separation of other ionic compounds. The methods originated in the field of solvent extraction. An ionised compound (A_{aq}^+) that is water soluble can be extracted into an organic solvent by using a suitable counter ion (B_{aq}^-) to form an ion-pair, according to the equation:

$$A_{aq}^+ \; + \; B_{aq}^- \; \rightleftharpoons \; (A^+ \; B^-)_{org}$$

The ion-pair $(A^+ \; B^-)$ behaves as if it is a nonionic polar molecule, soluble in organic solvents. By choosing a suitable pairing ion and adjusting its concentration, the ion A^+ can be efficiently extracted into an organic phase. Similarly, anions can be extracted by using a suitable cationic pairing ion. In ion-pair chromatography we use a reverse phase separation on a C-18 column, with the ion-pairing reagent added to the mobile phase. The ion-pairs are separated as neutral polar molecules.

There is still some debate about the mechanism in this method of separation. The simplest model assumes that ion-pairs are formed in the mobile phase and travel through the system as neutral species. Separation occurs by partitioning of these neutral ion-pairs between the mobile phase and the C-18. This mechanism cannot explain all the experimental results, and there is no doubt that it is a considerable over simplification. Recent ideas about the mechanism suggest that it involves a combination of partition and ion-exchange.

Typical ion-pairing reagents are, for cations, alkyl sulphonic acids, eg pentane, hexane, heptane or octane sulphonic acid, and for anions, tetrabutylammonium or dibutylamine ammonium salts. In ion-pair chromatography the retention of solutes can be controlled in a number of ways:

(a) By varying the chain length of the pairing agent. Retention increases as chain length increases.

(b) By varying the concentration of the pairing agent. Retention increases as the amount of the pairing agent increases.

(c) If we use a pH at which some solutes in our sample are ionised whilst others are not, then only the ionised solutes will be affected by any changes in the type or concentration of the pairing agent; we can alter the column selectivity for the ionised solutes without affecting the retention of the others.

(d) By changing the concentration of organic solvent in the mobile phase.

∏ How would a change in the concentration of organic solvent affect retention?

This will follow the rules for reverse phase separations (Fig. 3.1c). Increasing the amount of organic solvent will decrease the polarity of the mobile phase and will decrease retention.

Fig. 3.3e shows the separation of weak acids on a C-18 column with a mobile phase of methanol/water 50:50 + tetrabutylammonium phosphate. The pH of the mobile phase is about 7.5, as at this pH both weak acids are fully ionised and form ion-pairs with the pairing agent.

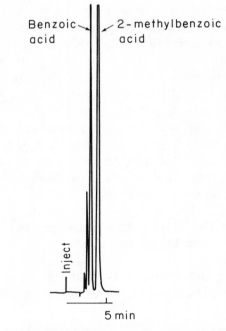

Fig. 3.3e.   *Ion pair separation of weak acids*

Column: 10 μm C-18 bonded phase, 30 cm × 4 mm
Mobile phase: CH_3OH/H_2O 50:50 + tetrabutyl ammonium phosphate, pH 7.5. Flow rate 2 cm^3 min^{-1}
Detector: Uv absorption, 254 nm

Better resolution is obtained here than in the ion-exchange separation of these two (Fig. 3.3c). The benzoic acid, being the more polar, is eluted first.

Fig. 3.3f(i) shows the use of a combination of ion-pairing and ion-suppression to separate a mixture of acids and bases. The pH of the mobile phase is about 2.5, as at this pH the maleic acid is unionised and elutes quickly as a very polar molecule on the reverse phase column. The other solutes are all weak bases which at pH 2.5 are fully protonated and pair with the pentane sulphonic acid anion.

The retention of the paired solutes can be increased by increasing the chain length of the pairing agent, as shown in Fig. 3.3f (*ii*).

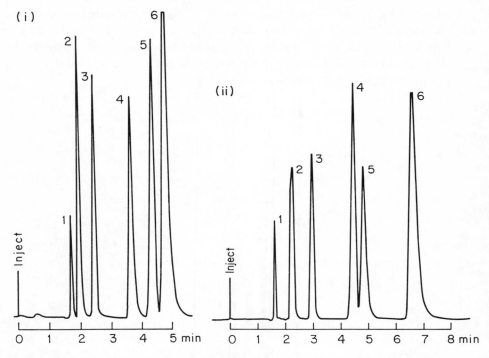

Fig. 3.3f. *Separation of a mixture of acids and bases by ion-pairing/ion-suppression*

(*i*)

Column:	10 μm C-18 bonded phase, 30 cm × 4 mm
Mobile phase:	CH₃OH/H₂O 50:50 both containing 0.005 mol dm⁻³ pentane sulphonic acid pH 2.5. Flow rate 2 cm³ min⁻¹
Detector:	Uv absorption, 254 nm
Sample:	1 = maleic acid 2 = phenylephrine 3 = norephedrine 4 = naphazoline 5 = phenacetin 6 = pyrilamine

(*ii*)

Conditions as Fig. 3.3f(*i*) except that hexane sulphonic acid is used as the ion-pair reagent.

Ion-pairing techniques have also been used in the separation of inorganic ions. Fig. 3.3g shows an example. Detection here is by indirect uv absorption. As the ions themselves do not absorb strongly,
an absorbing substance (potassium hydrogen phthalate) is added to
the mobile phase. The solute anions, being transparent at the detection wavelength, cause decreases in absorbance, ie negative peaks, as
they are eluted. Reversing the polarity of the chart recorder enable
positive peaks to be recorded.

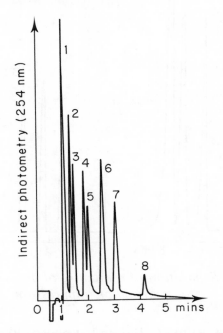

Fig. 3.3g. *Separation of inorganic ions using indirect photometric
detection*

Column:	5μm C-18 bonded phase, 25 cm × 4.6 mm
Mobile phase:	0.001 mol dm⁻³ tetrabutyl ammonium hydroxide + potassium hydrogen phthalate, pH8. Flow rate 1 cm³ min⁻¹
Detector:	Uv absorption, 254 nm

Column: $5\mu m$ C-18 bonded phase, 25 cm × 4.6 mm
Mobile phase: 0.001 mol dm^{-3} tetrabutyl ammonium hydroxide
                 + potassium hydrogen phthalate, pH8. Flow
                 rate 1 cm^3 min^{-1}
Detector: Uv absorption, 254 nm
Sample: 1 = chloride   2 = nitrite   3 = nitrate
                 4 = bromide   5 = phosphate   6 = phosphite
                 7 = sulphate   8 = iodide.

SAQ 3.3a Explain the method used in each of the following:

(*i*) Separation of aspirin and norephedrine (1-phenyl,2-aminopropanol)

Column: C-18; mobile phase: CH_3OH/H_2O 50:50 + heptane sulphonic acid (pH about 3.5).

(*ii*) Chromatography of 4-aminobenzoic acid

Column: C-18; mobile phase: CH_3OH/H_2O 50:50 + tetrabutylammonium hydroxide (pH about 7.5).

Summary

Ionic solutes can be separated by ion-exchange chromatography using microparticulate resins or bonded ion-exchangers based on microparticulate silica. Such separations are often achieved more easily by ion-suppression or ion-pairing techniques, which use bonded phase columns in the reverse phase mode.

Objectives

You should now be able to:

- describe the operation of an ion-exchange material in hplc;

- recognise the factors that affect solute retention times in ion-exchange chromatography;

- appreciate that ionic solutes can often be easily separated by ion-suppression or ion-pairing techniques;

- specify the experimental conditions that are used in ion-pair chromatography.

3.4. ADSORPTION AND EXCLUSION CHROMATOGRAPHY

3.4.1. Adsorption Chromatography

Unmodified silicas are used in this mode of hplc. The adsorption sites on the surface of silica are silanol (Si—OH) groups. These can be present as isolated groups, or can be hydrogen bonded to one another. With chromatographic silicas, the relative numbers of each type of silanol group depend on the type of silica and how it has been prepared and treated. The two types of silanol group have different adsorptive strength, and can be deactivated by moisture or by other polar solvents. Usually, for chromatography, the silica is activated by heating at 150–200 °C and then partly deactivated by the addition to the mobile phase of small amounts of water or some other polar organic solvent. This is done to try to standardise the activity of the adsorbent, but reproducibility is always a problem with chromatography on unmodified silica, and to obtain reproducible behaviour often requires lengthy conditioning procedures. With silica and a nonpolar mobile phase modified with a small amount of polar solvent it is possible that a layer of the polar component is adsorbed at the silica surface and that both adsorption and partition (between this layer and the mobile phase) are contributing to the separation.

The dissolved solute molecules X compete with mobile phase molecules S for a place on the adsorbent surface:

$$X + S_{ads} \rightleftharpoons X_{ads} + S$$

The strength of the interaction between the adsorbent and the solute molecules increases as the polarity of the solute increases. Thus we can increase retention of our solute molecules (X) by decreasing the polarity of the mobile phase (S), which will shift the equilibrium above to the right. Polar solute molecules are strongly held on unmodified silica and tail badly, so the method is useful only for solutes having low or medium polarity.

Fig. 3.4a shows the chromatogram of some phthalates on a silica column using ethyl ethanoate/*iso*-octane 5 : 95 as the mobile phase. Some of the peaks are identified.

∏   (*a*) In what order would you expect the other solutes to elute?

        (*b*) What changes would you expect in the chromatogram if the mobile phase was changed to butyl ethanoate/*iso*-octane 5 : 95?

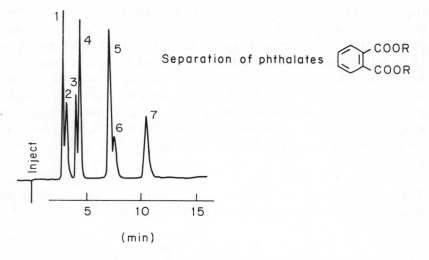

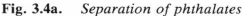

(min)

Fig. 3.4a.   *Separation of phthalates*

Column: 10 μm silica 30 cm × 4 mm
Mobile phase: ethyl ethanoate/*iso*-octane 5 : 95.
Flow rate: 2 cm³ min⁻¹
Detector: Uv absorption, 254 nm
Sample: R = CH₃, C₂H₅, C₆H₅ (6), *n*-C₄H₉,
                     iso-C₄H₉ (3), *iso*-C₈H₁₇ (2), *n*-C₈H₁₇

(*a*) In a normal phase separation like this one the solutes
    elute in order of increasing polarity, so the order of elu-
    tion would be, first, dioctyl followed by dibutyl, diethyl,
    dimethyl. You should compare this chromatogram with
    Fig. 4.3k which involves the same kind of sample but on
    a reverse phase system. The order of elution is reversed
    with the most polar solute eluting first:

(*b*) The mobile phase is now a little less polar, so we would
    expect the retention of all the solutes to increase. The
    chromatogram is shown in Fig. 3.4b. Changing the mo-
    bile phase will also change the selectivity and you can
    see that this has happened if you look at the diethyl
    and diphenyl peaks, where the order of elution has been
    reversed.

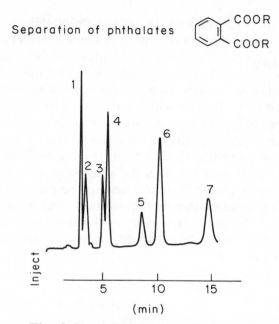

Fig. 3.4b. *Separation of phthalates*

Column:	10 μm silica 30 cm $\times$ 4 mm
Mobile phase:	butyl ethanoate/*iso*-octane 5 : 95
Detector:	Uv absorption, 254 nm
Sample:	1 R = n-C_8H_{17} 2 *iso*-C_8H_{17} 3 *iso*-C_4H_9
	4 n-C_4H_9 5 C_6H_5 6 C_2H_5 7 CH_3

3.4.2. Exclusion Chromatography

Exclusion chromatography is a technique for separating molecules based on their effective size and shape in solution. The technique is often called gel permeation chromatography if used with organic solvents or gel filtration if used with aqueous solvents.

The stationary phases used in exclusion chromatography are porous particles with a closely controlled pore size. Unlike other chromatographic modes, in exclusion chromatography there should be no interaction between the solute and the surface of the stationary phase.

Depending on their size and shape, solute molecules may be able to enter the pores of the stationary phase particles. Molecules comparable in size to the mobile phase molecules will be able to diffuse throughout the entire porous network. Larger molecules may be excluded from the narrower parts of the porous structure but will be able to move freely in the wider passages. The larger the solute molecule, the fewer places in the porous structure it will find that it can get into. Finally, there may be solute molecules that are so large that they are completely excluded from the pores. These excluded molecules can travel only through the relatively wide channels between the stationary phase particles, and so are eluted rapidly from the column. The smaller the molecule, the more easily it will be able to penetrate the pore structures of the stationary phase particles, and the longer it will be retained on the column. The process is illustrated in Fig. 3.4c.

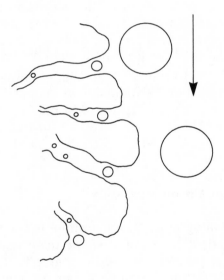

Fig. 3.4c.   *Separation by exclusion*

Large molecules are excluded from the internal pores of the stationary phase, and therefore travel in the channels between the stationary phase particles. Smaller molecules that can enter the porous network travel more slowly.

In exclusion chromatography, the total volume of mobile phase in the column is the sum of the volume external to the stationary phase particles (the void volume, V_o) and the volume within the pores of the particles (the interstitial volume, V_i). Large molecules that are excluded from the pores must have a retention volume V_o, small molecules that can completely permeate the porous network will have a retention volume of $(V_o + V_i)$. Molecules of intermediate size that can enter some, but not all of the pore space will have a retention volume between V_o and $(V_o + V_i)$. Provided that exclusion is the only separation mechanism (ie no adsorption, partition or ion-exchange), the entire sample must elute between these two volume limits.

In Fig. 3.4d the relative molecular mass of the solute, M_r, is plotted on a log scale against the retention volume. The interstitial volume, which represents the volume range within which separations occur, and the size range of solutes that are eluted in this volume range, depend on the sort of material that is used for the stationary phase. Because for a given separation, V_o and V_i are constant, we can reliably predict the total volume of solvent or the time taken for a particular analysis. The calibration curve is established by determining the retention volume for standards of known M_r.

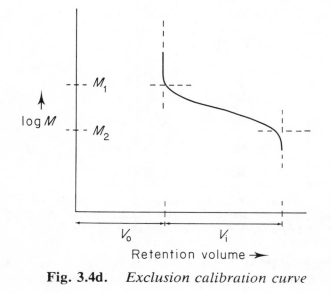

Fig. 3.4d. *Exclusion calibration curve*

Molecules with relative molecular mass $> M_1$, are totally excluded
from the stationary phase and have retention volume V_0. Molecules
with relative molecular mass $< M_2$ totally permeate the stationary
phase and elute at $(V_0 + V_i)$. Molecular sizes in between these two
partly permeate the stationary phase and elute between V_0 and $(V_0 + V_i)$.

The materials originally used for exclusion chromatography were
semirigid gels of cross-linked dextran (a carbohydrate) or polyacry-
lamide. These cannot withstand the high pressures used in hplc.
Modern stationary phases used in this technique are microparticulate materials consisting of styrene–divinylbenzene copolymers, silica, or porous glass. They are available in a range of pore sizes for
the separation of different M_r ranges. The styrene–divinylbenzene
types are used with organic solvents, as aqueous solvents often cause
excessive shrinkage of the column bed, producing voids and channels which lead to a loss of efficiency. Fig. 3.4e shows the available
sizes and M_r ranges of a styrene-divinylbenzene exclusion stationary
phase (Ultrastyragel, manufactured by Waters). It has a particle size
of 10 μm and produces efficiencies of around 46 000 plates m^{-1}.
Apart from the 100 Å type, they can be used at temperatures up to
145 °C.

Pore size, Å	M_r range
100	50–1500
500	100–10^4
10^3	200–3 $\times$ 10^4
10^4	5 $\times$ 10^3–6 $\times$ 10^5
10^5	5 $\times$ 10^4–4 $\times$ 10^6
10^6	2 $\times$ 10^5–10^7

Fig. 3.4e. *Styrene–divinylbenzene exclusion stationary phase
(Ultrastyragel, manufactured by Waters)*

The pore size of these is given in Angstroms (1 Å = 10^{-8} cm)
and relates to the chain length of a polystyrene molecule just large

enough to be totally excluded from all the pores of the gel. The mass range is the range of M_r (determined using polystyrene standards) that is partially excluded. To provide partial exclusion over a wide range of M_r, a number of columns can be used in series, each column containing a different molecular size range.

∏   A 7.8 mm × 30 cm column containing the 10^4 Å stationary phase above is used with methylbenzene as the mobile phase at a flow rate of 1.1 cm^3 min^{-1}. A sample of polystyrene standards dissolved in benzene is injected. The standards have molecular masses of 775 000, 442 000, 6200 and 2800. The void volume of the column is 6 cm^3 and the interstitial volume is 5 cm^3.

(*a*) Which is the first solute to elute, and what is its retention volume,

(*b*) Which of the solutes are partially excluded,

(*c*) How long does the separation take?

(*a*) The 775K standard will be totally excluded, and will elute at $V_O = 6$ cm^3,

(*b*) The 442K and the 6.2K standards will be partially excluded,

(*c*) The 2.8K standard and the benzene will totally permeate the column and will elute together at $(V_O + V_i) = 11$ cm^3. This takes $11/1.1 = 10$ min.

Silica stationary phases for exclusion can be used with either organic or aqueous solvents. Some types are bonded phases, others are unmodified. When aqueous phases are used with silica exclusion columns, small amounts of polar mobile phase modifiers (inorganic salts or polar organic solvents) often have to be used to reduce adsorption effects.

The choice of mobile phase for exclusion chromatography is simpler than for other hplc modes, as only one solvent is required. For polymers, solubility considerations often govern the choice of mobile phase. Tetrahydrofuran or chlorinated hydrocarbons are often used for polymers that are soluble at room temperature. Some polymers (eg polyethene) require temperatures of around 150 °C for solution; for these, di- or trichlorobenzenes can be used as the mobile phase.

Exclusion chromatography was originally used mainly for the characterisation of polymers, although it now has many uses in other fields. Synthetic polymers have a range of molecular size, and the distribution of M_r will affect many of the important bulk properties of the polymer such as hardness, brittleness, tensile strength and so on. Small changes in M_r distribution may cause large variations in the end-use performance of a polymer. Before the advent of exclusion chromatography, the determination of M_r distribution in polymers was commonly done by fractional precipitation methods, which were difficult, lengthy and inaccurate. These days, the mass distribution and the average M_r of the polymer can easily be calculated from the exclusion chromatogram.

Exclusion chromatography is also useful in the separation of small molecules from interfering matrices of larger molecules, for example in foods or other samples of biological origin. It can be used as the first step in the sequential analysis of complex unknown organic mixtures, which are first separated on a size basis by exclusion, then the collected fractions can be further separated by normal or reverse phase chromatography, where the separation is based on chemical differences.

Fig. 3.4f shows an exclusion chromatogram on unmodified silica. The sample is an epoxy resin with an average M_r of 900. Fig. 3.4g shows the determination of pesticide residues in a sample of chicken fat, and is an example of how exclusion can be used to clean up complex samples. First, a pesticide-free sample of the fat is run as a blank, then the blank is spiked with the pesticides to determine their retention volumes. When the sample is injected, the eluent containing the pesticides is collected. The solvent is evaporated, the residue dissolved in acetonitrile and the pesticides are then separated on a reverse phase column.

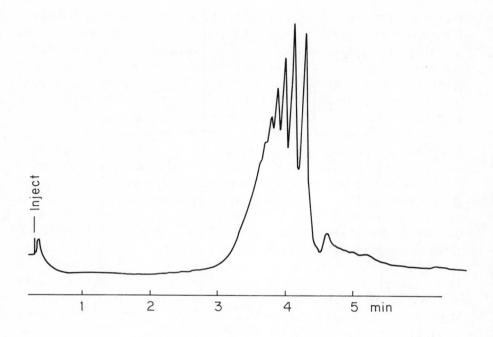

Fig. 3.4f. *Exclusion chromatogram of an epoxy resin*

Column: 5 μm silica 25 cm × 4.9 mm
Mobile phase: tetrahydrofuran + 1% H_2O. Flow rate 1 cm^3 min^{-1}
Detector: Uv absorption, 254 nm
Sample: 1 μl Epikote 1001 in tetrahydrofuran

Epikote 1001 is a synthetic epoxy resin with an average relative molecular mass of 900

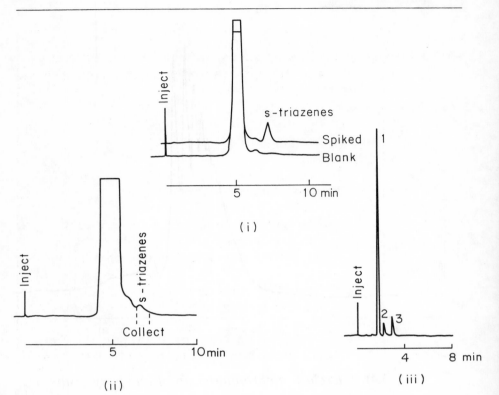

Fig. 3.4g.   *Clean up of a sample of chicken fat containing pesticides,*
using exclusion chromatography

(*i*)   Fat sample, blank and spiked with pesticides
(*ii*) Fat sample containing pesticides

    Column: 100 Å μ-styragel (styrene-divinyl benzene
                       microparticulate resin), 122 cm × 7.8 mm
    Mobile phase: trichloromethane, flow rate 2 cm^3 min^{-1}
    Sample: 100 μl chicken fat in trichloromethane

(*iii*) Reverse phase separation of pesticide residues.

    Column: 10 μm C−18, 30 cm × 4 mm
    Mobile phase: CH_3CN/H_2O 60 : 40, flow rate 2 cm^3 min^{-1}.
    Detector: Uv absorption, 254 nm.
    Peaks: 1 simazine   2 atrazine   3 propazine

SAQ 3.4a

The following statements refer to the different modes of hplc. Indicate whether the statements are true (T) or false (F).

(*i*) In adsorption chromatography, a non-polar mobile phase is used.

(*ii*) Polar molecules can easily be separated by adsorption chromatography.

(*iii*) The retention times of solutes in exclusion chromatography can be altered by changing the polarity of the mobile phase.

(*iv*) Exclusion chromatography is useful only for the separation of large molecules.

(*v*) In reverse phase chromatography, the mobile phase is more polar than the stationary phase.

(*vi*) In reverse phase chromatography using bonded silica packings, the bonded group is non-polar.

SAQ 3.4b   Which mode of hplc would you choose for each
                of the following:

                (*i*)   Identification of plasticisers in poly-
                        chloroethene (polyvinyl chloride). Com-
                        mon pvc plasticisers are dibutyl, dioctyl
                        and dinonyl phthalates.

                (*ii*) Separation of tranquillisers.

                        Structures:

                        Place SAQ 3.4b a/w (*i*) here

                        Diazepam (valium): $R_1 = -CH_3$,
                        $R_2 = -H$

                        Oxazepam (serax): $R_1 = -H$, $R_2 = -OH$

                (*iii*) Separation of a mixture of synthetic food
                        dyes.

                        Structures:

                        Place SAQ 3.4b(*ii*) a/w here

                        Amaranth: $R_1 = R_2 = -SO_3Na$,
                                     $R_3 = -H$

                                                                $\longrightarrow$

SAQ 3.4b (cont.)

Ponceau 4 R: $R_1 = -H,$
$R_2 = R_3 = -SO_3Na$

Summary

Adsorption chromatography uses unmodified silica with a relatively nonpolar mobile phase. It is used for the separation of solutes of relatively low polarity, although such separations are now often achieved more easily on bonded phases.

Exclusion chromatography separates solutes that differ in size and shape. The technique is used extensively in the investigation of macromolecules and in the separation of small molecules from an interfering matrix of larger molecules.

Objectives

You should now be able to:

- predict the order of elution in simple cases using adsorption chromatography on unmodified silica;

- appreciate the difficulty of obtaining reproducible behaviour in adsorption chromatography;

- identify stationary and mobile phases used in exclusion chromatography;

- understand the mechanism by which solutes are separated in exclusion chromatography;

- recognise the areas in which exclusion chromatography is used.

4.   Some Applications of hplc

In this section we are going to look at some case studies to see how hplc experimental methods are developed. I am not going to give a long list of applications, because these are easy to find elsewhere, and sometimes do not make very interesting reading. Most textbooks on hplc have lists of applications, eg the book by Hamilton and Sewell (2nd Edn, Chapter 8), and applications can also be obtained from a number of journals (eg Analytical Chemistry annual reviews).

4.1.   FACTORS AFFECTING RESOLUTION

What we are looking for in a chromatographic separation is to achieve satisfactory resolution of our solutes in the minimum amount of time. Resolution (R_S) describes the degree of separation of one component from another, and is defined as the difference in retention volumes of the two solutes divided by their average peak width.

$$R_S = (V_{R2} - V_{R1})/0.5 \times (w_1 + w_2) \qquad (4.1a)$$

Fig. 4.1a shows how resolution can be measured from a chromatogram. The values of V and w can be measured in volume, time or chart length (from the point of injection) as long as we use the same units for each of them. When two peaks are just resolved to the baseline, this corresponds to a resolution of 1.5.

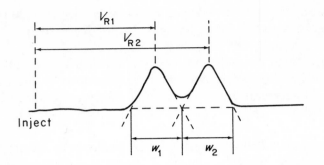

Fig. 4.1a.   *Measurement of resolution*

A peak on a chromatogram is usually identified by some measure of retention, for example retention time or chart length where the length or time is measured from the injection point to the peak apex. A useful quantity to locate or identify a peak is the capacity factor (k') which is defined as follows:

$$k' = (V_R - V_o)/V_o \qquad (4.1b)$$

V_R = the retention volume (time or length) of our solute

V_o = the retention volume (time or length) of an unretained solute.

In terms of volume, V_o is a measure of the system dead volume from the injector to the detector. For a well designed system with low extra-column dispersion, V_o will be roughly equal to the dead volume of the column, that is, the volume of the column not occupied by the packing particles.

You can see from Eq. 4.1b that k' is simply a measure of retention in terms of V_o and V_R so once these are identified on a chromatogram, it is easy to calculate k' values.

The capacity factor tells us where the peaks elute relative to V_o. The separation factor (α) tells us where the peaks elute relative to each other. It is defined, for two peaks, as the ratio of the capacity factors, with the larger one in the numerator.

$$\alpha = k_2'/k_1' = (V_{R2} - V_o)/V_{R1} - V_o) \qquad (4.1c)$$

A separation factor of one means that only one peak will be produced from our two solutes, ie the value of R_S is zero.

Neither the capacity factor nor the separation factor take into account the effect of dispersion, which is measured by the plate number or the plate height of the column. These were defined in Section 2.3.2.

If we assume that the width of the two peaks is the same, then it is possible to show that the relation between these quantities is:

$$R_S = 0.25[(\alpha - 1)/\alpha] \times [k_2'/(1 + k_2')] \times N_2^{\frac{1}{2}} \qquad (4.1d)$$

Where k_2' and N_2 refer to the second of the two components.

This shows that for a desired degree of resolution three conditions have to be met: (*a*) the peaks have to be retained on the column ($k_2' > 0$), (*b*) the peaks have to be separated from each other ($\alpha > 1$), and (*c*) the column must develop some minimum number of plates.

If we can increase k_2', other things being equal, the resolution between our two peaks will increase significantly at first, but the effect will diminish at higher values of k_2'. You can show this easily by plotting a few values of $k_2'/(1 + k_2')$ against k_2'. The optimum range for k_2' depends on the efficiency of the column that is used. For a reasonable compromise between resolution and the time of the separation, we want k_2' for our peaks to lie between about 1 and 10. Eq. 4.1d also shows that we can increase resolution by increasing efficiency, although to double the resolution we have to increase the efficiency by a factor of four. Efficiency can usually be increased by operating at a lower flow rate (which increases the time of the analysis), by increasing the length of the column or reducing the particle size of the packing (which increases the pressure drop across the column).

Fig. 4.1b shows what happens to two partly resolved peaks when N_2, α or k_2' are changed. If we increase N_2, other things being equal, the solutes appear at the same places on the chromatogram but the resolution is better because each peak width has decreased. Increasing

k_2' improves resolution by causing the solutes to spend more time in the stationary phase. Selectivity can be altered by changing the nature of the stationary phase (eg for a reverse phase separation using non-polar bonded silica there are a variety of stationary phases, including phenyl, C-8 or C-18), or by changing the nature or the composition of the mobile phase (see Section 3.1.3).

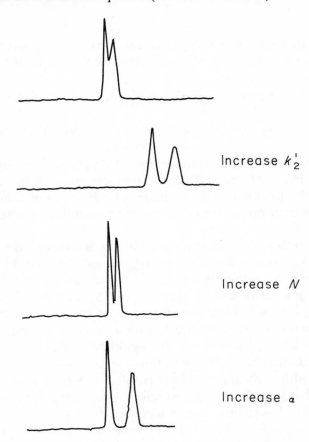

Increase k_2'

Increase N

Increase α

Fig. 4.1b. *Effect of N_2, α and k_2' on resolution*

In practice, if we change α or k_2' then both of the other two (k_2' and N_2 or α and N_2) will change as well, so that there is usually a certain amount of trial and error involved in developing an hplc method, as the following examples will show.

SAQ 4.1a	The chromatogram in Fig. 4.1c has some partly resolved peaks. Assume that the first peak is an unretained solute and take V_o as the position where the first peak starts to elute.

(*i*) Draw a scale on the abscissa axis of the chromatogram marking k' values from 0 to 5.

(*ii*) Determine k' for each peak on the chromatogram.

(*iii*) Measure peak widths and calculate the resolution (R_S) between peaks 1 and 2, peaks 2 and 3 and peaks 4 and 5. For the peaks that are partly resolved you will have to extrapolate the linear part of the side of each peak down to the baseline, as in Fig. 4.1a.

(*iv*) Calculate the selectivity (α) for peaks 2 and 3, 3 and 4, 4 and 5.

(*v*) For the last two peaks, find the plate number and plate height of the column using Eq. 2.3c (the column is 25 cm long and has a relatively low efficiency).

(*vi*) Tabulate the results.

SAQ 4.1a

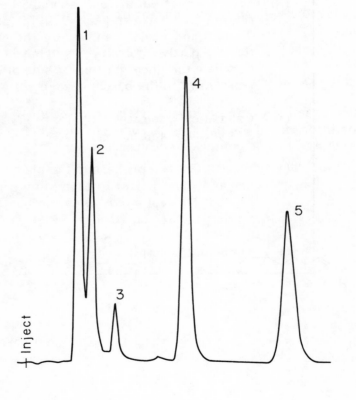

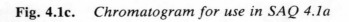

Fig. 4.1c.   *Chromatogram for use in SAQ 4.1a*

Summary

The object of a chromatographic separation is to achieve satisfactory resolution of solutes in the minimum time. Resolution is influenced by the capacity factor of the solutes and the selectivity and plate number of the column.

Objectives

You should now be able to:

- define resolution, capacity factor and selectivity;

- calculate these quantities from measurements on peaks in a chromatogram.

4.2.   DEVELOPING A SEPARATION

The first example demonstrates the steps involved in working out the conditions for the separation of some steroids. Fig. 4.2a shows the structure of some conjugated oestrogens.

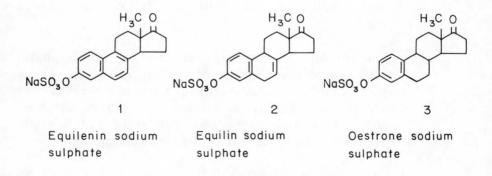

1	2	3
Equilenin sodium sulphate	Equilin sodium sulphate	Oestrone sodium sulphate

Fig. 4.2a.   *Structure of conjugated oestrogens*

The problem is that we want to separate these from one another, and from excipients in a commercial tablet. To get an idea of the conditions needed for the separation, you have to concentrate on the differences between them.

∏   Look at the structures and see if you can suggest:

(*a*) What sort of column packing is needed for the separation

(*b*) What sort of mobile phase should be used with the packing that you choose,

(*c*) Whether or not these compounds would be soluble in this mobile phase,

(*d*) Which detector would be the most suitable.

(*a*) If you suggested any sort of a non-polar packing then you are thinking along the right lines. The differences between these structures are in the nonpolar parts of the molecules, so we need a nonpolar packing to exploit these differences; ideally a packing that is very similar chemically to the parts of the molecules that differ. A phenyl bonded phase would probably be the best bet, but in this case a non-polar C-18(ODS) column was used.

(*b*) and (*c*) An C-18 column needs a polar mobile phase, such as a methanol–water or acetonitrile–water, so as a starting point a 50 : 50 methanol/water mixture was chosen. Because the three compounds are sodium salts, they should be soluble in this solvent mixture. This is easily checked using standards of the three compounds.

(*d*) The aromatic rings suggest that uv absorption would be a suitable method of detection.

Sample preparation consists of crushing some tablets, mixing with 50:50 methanol/water, diluting to the mark in a volumetric flask and then filtering off any insoluble excipients. We are now ready to go, and Fig. 4.2b (*i*) shows the results of the first injection.

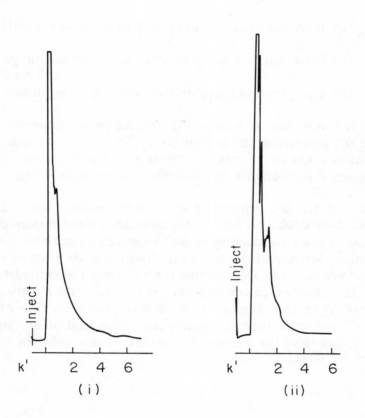

Fig. 4.2b. *Chromatogram of a steroid tablet*

Mobile phase:	(*i*) CH_3OH/H_2O 50:50
	(*ii*) CH_3OH/H_2O 35:65
Column:	10 μm C-18 bonded phase 30 cm × 4 mm
Flow rate:	2 cm^3 min^{-1}
Detector:	Uv absorption, 254 nm

This is a disaster. There is little or no separation, and almost everything is eluting from the column immediately at a k' little more than zero. The first step is to increase the retention times of all the solutes, ie increase the k' values.

∏   Would you do this by:

(*a*) increasing the amount of methanol in the mobile phase,

(*b*) increasing the amount of water in the mobile phase, or

(*c*) changing to an acetonitrile/water mobile phase?

The way to increase k' values with reverse phase columns is to increase the polarity of the mobile phase, so we want to increase the amount of water in the mixture rather than the methanol. It would not be sensible to change to acetonitrile/water at this stage.

In Fig. 4.2b (*ii*) the amount of water in the mobile phase has been increased to 65%. This chromatogram is not much better than the first one, but we are starting to get longer retention times and some resolution. We need to know if the things that are starting to separate out are the oestrogens or just rubbish from the excipients in the table. The next two chromatograms, in Fig. 4.2c, are for mixed standards using mobile phases containing, respectively, 65% and 80% water. These show that the oestrogens are indeed being retained a little longer than the excipients, so that we are going in the right direction.

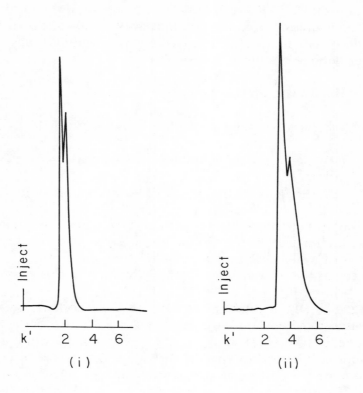

Fig. 4.2c. *Chromatograms of steroid standards*

Mobile phase: (*i*) CH_3OH/H_2O 35 : 65
(*ii*) CH_3OH/H_2O 20 : 80

The next chromatogram, Fig. 4.2d(*i*) is an injection of the tablet solution with the 80% water mobile phase. You can see from this that the oestrogens have k' values between 3 and 5 and that they are separated from the excipients. Although things are getting better, there are still several problems left to solve.

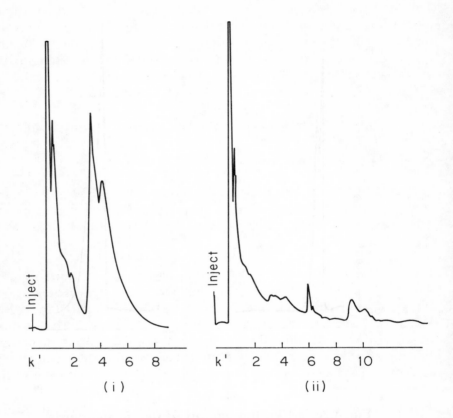

Fig. 4.2d.   *Chromatograms of steroid tablet*

Mobile phase: (*i*) CH_3OH/H_2O 20 : 80
(*ii*) $CH_3OH/0.001$ mol dm^{-3} KH_2PO_4
20 : 80 (pH5)

∏   Can you see two undesirable features in the chromatogram?

The two undesirable features are:

(*a*) The oestrogens are not properly resolved (there are only two peaks)

(*b*) the oestrogen peaks are tailing.

The tailing is probably caused by a mixed mechanism, for instance adsorption on active silica sites that are not end-capped. To reduce this, we can try adding a salt to the water. To get better resolution we need to change the selectivity, α, which means changing the chemistry of the mobile phase, or increasing the plate number of the column, or both.

ΠΠ   Do you think that the addition of a salt as a modifier will affect k', α or N (or a combination of them) and if so, what will happen?

Addition of a salt will have no effect on N but will have a drastic effect on the k' values as we are increasing the polarity of the mobile phase. We would expect k' values to increase.

Because the addition of a salt will change both k' and quite likely α values we need to try this first. Fig. 4.2d (*ii*) shows the chromatogram obtained using 0.001 mol dm^{-3} KH$_2$PO$_4$ (pH 5) instead of the water. What has happened here is that the oestrogens either do not elute at all, or if they do elute it takes far too long. This is not surprising, because we have made the mobile phase a great deal more polar and forced the oestrogens to interact more strongly with the C-18.

It is tempting at this point to say that we made a mistake in adding the phosphate, but in fact there is no way of knowing the effect the phosphate has had on tailing and selectivity until we can get the k' values back into the 1–10 region and have a look at the peaks. So we now need to decrease the k' values, which means changing the water–organic ratio in the mobile phase again. Fig. 4.2e shows the chromatogram obtained using phosphate, but with the original 50 : 50 methanol/water ratio. The k' values are in the required range, the tailing problem has been eliminated, and the selectivity is better.

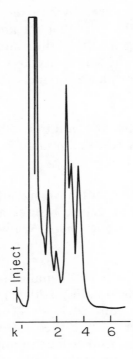

Fig. 4.2e. *Chromatogram of steroid tablets*

Mobile phase: $CH_3OH/0.001$ mol dm^{-3} KH_2PO_4 50:50

The resolution of the three oestrogens still has to be improved, so to proceed further we can either work on the selectivity, α, by using, instead of methanol, a different water-soluble solvent such as acetonitrile, tetrahydrofuran or dioxane, or we can try to improve the separation by increasing the plate number of the column. If we change the solvent, we cannot be sure about what will happen to the selectivity, and we may have to do a lot more experimental work to get any improvement. Increasing the plate number, if it can be done, is the easier of the two options. Fig. 4.2f shows the improvement that results when two 30 cm columns are used in series, with a flow rate of 1 cm^3 min^{-1}. The oestrogens are separated from the excipients and are also separated reasonably well from one another. The separation is complete in about 20 minutes.

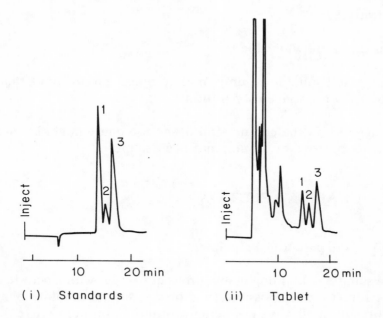

Fig. 4.2f. *Chromatograms of steroids*

Column: Two 30 cm × 4 mm columns containing 10 μm C-18 bonded phase
Flow rate: 1 cm^3 min^{-1}
Mobile phase: CH$_3$OH/0.001 mol dm^{-3} KH$_2$PO$_4$ 50:50 pH = 5
Detector: Uv absorption, 254 nm

Summary

Experimental conditions are developed for the separation of a sample of steroids in a commercial tablet.

Objectives

You should now be able to:

● suggest a suitable column, mobile phase and detector that might be used for a given separation;

● suggest how the column and/or mobile phase might be altered so as to improve resolution and peak shape.

4.3.   GRADIENT ELUTION

4.3.1.   Purpose and Principles

In a sample containing many different solutes, with isocratic elution it is sometimes impossible to choose a suitable mobile phase that will result in all k' values being within the optimum range. If this is the case, the chromatogram may appear as in Fig. 4.3a.

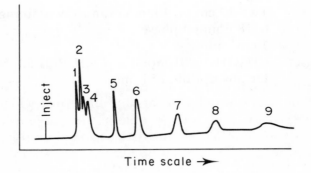

Fig. 4.3a.   *Chromatogram under isocratic conditions*

The early peaks appear at k' values between 0 and 1 and are poorly resolved. Peaks 5 and 6 are well resolved, but peak 7 and subsequent peaks are getting very dispersed, and are taking a long time to elute.

To improve the chromatogram we need to increase the k' values of the early peaks by using a mobile phase of lower eluting power, which will cause the solutes to spend more time in the stationary phase; and we also need to decrease the k' values of peaks 7–9 by using a mobile phase of greater eluting power in this region of the chromatogram.

The problem can be solved using gradient elution, where the composition of the mobile phase is altered during the separation, usually by blending two solvents with differing eluting powers in continually changing proportions. The shape of the gradient can be linear, convex or concave, as shown in Fig. 4.3b.

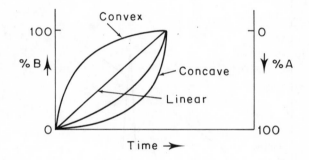

Fig. 4.3b.   *Possible gradient profiles showing the blending of solvent B with solvent A*

Many modern solvent delivery systems allow you to select a wide range of gradient profiles and to vary the time over which the gradient is delivered. As well as the profile of the gradient we have to consider a number of other factors, such as compatibility of the two solvents with the detector, or miscibility with the sample solution. For instance, if a uv absorbance detector is used and the absorbance of the two solvents are slightly different, then the baseline will drift during the gradient. This can be overcome by adding an unretained absorbing substance to one of the solvents to adjust them to a constant absorbance.

It is always important to run a blank gradient, ie a record of the detector response during the generation of a gradient. Fig. 4.3c shows

a linear blank gradient, from 100% water to 100% acetonitrile, run
on a C-18 column. The peaks in the chromatogram are artifacts from
the distilled water. What is happening is that the ODS column con-
centrates traces of organic compounds from the water at the head
of the column. As the proportion of organic solvent in the gradient
is increased, the organic material is eluted. Clearly, in this case the
water needs to be purified, but if we had not run a blank gradient
we would not have known whether the peaks came from the sample
or the mobile phase.

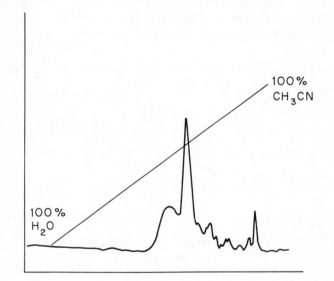

Fig. 4.3c. *Blank gradient (Record of detector response during the
generation of a gradient)*

     Column: 10 μm C-18 bonded phase 30 cm × 4 mm
     Solvent: A = H_2O B = CH_3CN 0%B → 100%B
     Flow rate: 0.5 cm^3 min^{-1}
     Detector: Uv absorption 254 nm

Fig. 4.3d and 4.3e show some of the other problems that can occur
with two solvent gradients.

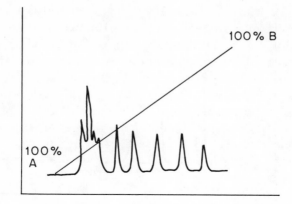

Fig. 4.3d. *Chromatogram using a two solvent gradient*

∏   What is the problem with the chromatogram in Fig. 4.3d?

The resolution is poor at the start of the chromatogram.

∏   Would you improve the resolution by:

(*a*) Keeping the same gradient shape but decreasing the strength (eluting power) of solvent A

(*b*) Keeping A and B the same but increasing the time over which the gradient is delivered

(*c*) Increasing the strength of solvent B

(*d*) Using a concave gradient?

Option (*c*) would make things worse, but any of the others should improve matters. I would try option (*a*) first. The problem is caused by A being too strong, or B being added too quickly. Use of a weaker solvent for A should increase the retention times of solutes 1–4 and increase their k' values. Addition of B more slowly, by increasing the time of the gradient, delaying the start of the gradient, or using a concave gradient, should also produce an improvement.

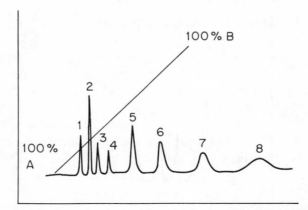

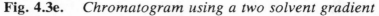

Fig. 4.3e.   *Chromatogram using a two solvent gradient*

∏   What is the problem with the chromatogram in Fig. 4.3e?

The second solvent is too weak, so that peaks continue to appear after the end of the gradient (peaks 7 and 8 are eluted isocratically and are highly dispersed).

∏   See if you can predict the appearance of the chromatogram if both A was too strong and B was too weak.

If this was the case then the chromatogram would resemble the isocratic chromatogram shown in Fig. 4.3a, with poorly resolved peaks at the start and highly dispersed peaks at the end.

If solvent B is too strong, then the middle or later peaks may be poorly resolved. In extreme cases of this, the solvent B may itself be strongly held at the top of the column (the column is now separating the solvent mixture, the effect being called solvent demixing). If this occurs then pure A will pass through the column until the column is saturated with solvent B.

The use of gradients is not without disadvantages, as solvent delivery systems are expensive and analysis times are relatively long, because the gradient has to be retraced between samples. Flow programming, in which the flow rate of solvent is caused to change during the separation, is sometimes used as an alternative to gradient elution.

4.3.2. Development of a Gradient Profile

This Section describes the practical development of a gradient profile, using as a sample Triton X-100, which is a non-ionic surface active agent. Chemically, it is a polyethene glycol which has a range of relative molecular mass (the average M_r is 628). We want to resolve the different molecular sizes by a normal phase separation using a silica column. The mobile phase is trichloromethane containing a small amount of dimethyl sulphoxide (dmso).

Fig. 4.3f shows the chromatogram obtained with a 3:97 dmso/trichloromethane mobile phase.

Fig. 4.3f. *Chromatogram of Triton X-100. Isocratic elution*

Column:	10 μm silica, 30 cm × 4 mm
Mobile phase:	dimethylsulphoxide/trichloromethane 3:97
Flow rate:	3 cm^3 min^{-1}
Detector:	Uv absorption, 280 nm

As in the previous example, we have little or no separation, and everything is eluting much too quickly.

∏   To improve matters, would you change the mobile phase to:

   (*a*) dioxane-dmso
   (*b*) dichloromethane/dmso
   (*c*) heptane/trichloromethane/dmso
   (*d*) heptane/dmso?

   To increase retention in a normal phase separation we need a less polar mobile phase, so option (*a*) would make things worse. All the other mixtures are less polar than the starting mobile phase, but mixture (*b*) only slightly less so, which would probably not make much difference. Heptane is non-polar, but the highly polar dmso is not soluble in it. It is best to keep the trichloromethane/dmso and add a non-polar solvent as a modifier as in (*c*). We can then change the polarity as we wish by altering the relative amounts of heptane and trichloromethane in the mixture.

The next three chromatograms, Fig. 4.3g, show the effect of increasing the amount of heptane in the mobile phase. The following mobile phases are used in the three chromatograms:

(*i*)   dmso/trichloromethane/heptane 3 : 40 : 60

(*ii*)   dmso/trichloromethane/heptane 3 : 20 : 80

(*iii*) dmso/trichloromethane/heptane 3 : 10 : 87

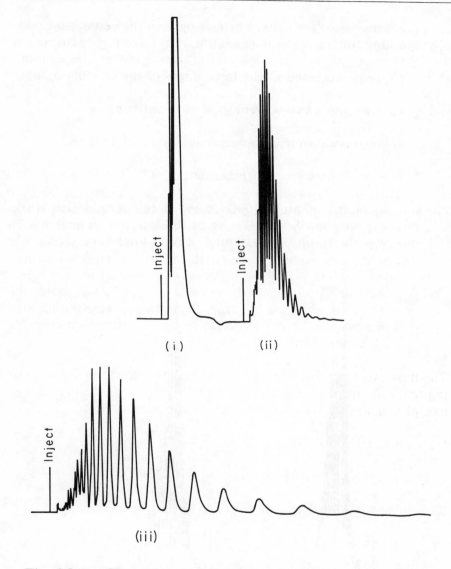

Fig. 4.3g. *Chromatograms of Triton X-100. Isocratic elution*

∏ In Fig. 4.3g(*iii*), what is wrong with the chromatogram, and how would you improve it by using a gradient?

The resolution is good for most of the chromatogram, but

compared to the other chromatograms the separation takes
a long time and the later eluting peaks are highly dispersed.
We need a gradient in which the eluting power of the mobile
phase is increased in the later stages of the chromatogram.

In Fig. 4.3h we use a two solvent gradient consisting of:

Solvent A dmso/trichloromethane/heptane 3 : 10 : 87

Solvent B dmso/trichloromethane 3 : 97

The starting mobile phase is 100% A and the linear gradient is run
to a final proportion of 50% B. Use of the longer gradient in 4.3h
(*ii*) improves the resolution, although it is still not very good.

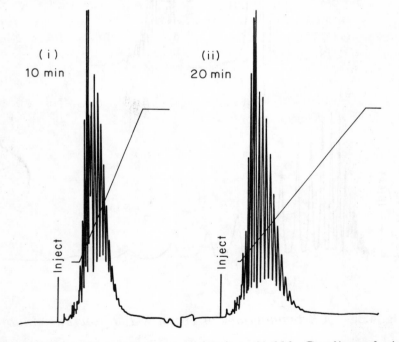

Fig. 4.3h. *Chromatograms of Triton X-100. Gradient elution*

Solvent A: dmso/trichloromethane/heptane 3 : 10 : 87
Solvent B: dmso/trichloromethane 3 : 97
Gradient: 0–50%B

∏ To improve resolution, would you:

(*a*) increase the time over which the gradient is delivered,

(*b*) use a solvent for B of lower eluting power (eg another mixture of dmso, trichloromethane and heptane),

(*c*) keep the same solvents but finish up with a smaller proportion of B, or

(*d*) use a concave gradient, so that B is added more slowly?

Any of these should produce some improvement. Option (*a*) seems the least promising, as only a small amount of the gradient is actually being used. Of the remainder, choices (*c*) and (*d*) are easier experimentally, as we can do these by reprogramming the gradient former, whereas if we change the mobile phase we have to spend time flushing the system through.

Fig. 4.3i shows the effect of changing the final proportion of B from 50% to 30% and finally to 20%.

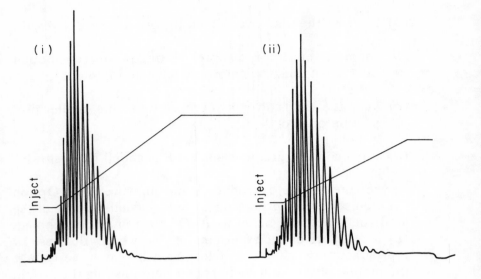

Fig. 4.3i.   *Chromatograms of Triton X-100. Gradient elution*

Solvent A:	dmso/trichloromethane/heptane	$3:10:87$
Solvent B:	dmso/trichloromethane	$3:97$
Gradients	(*i*) 0–30% B, 20 min	
	(*ii*) 0–20% B, 20 min	

At the smaller proportions of B, the resolution, although better, can still be improved, so finally in Fig. 4.3j (*i*) and (*ii*) we try adding B more slowly at the start of the run by using an exponential gradient and increasing the time. The peaks are now eluted over the entire range of the gradient, and most of the peaks are resolved almost to the baseline.

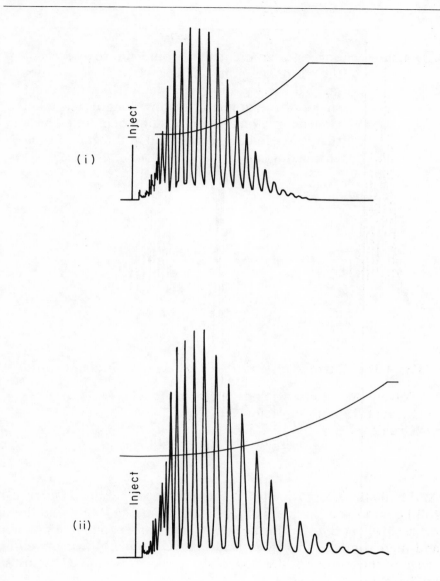

Fig. 4.3j.   *Chromatograms of Triton X-100. Gradient elution*

Solvent A: dmso/trichloromethane/heptane   3 : 10 : 87
Solvent B: dmso/trichloromethane   3 : 97
Exponential Gradients: (*i*) 0–20% B, 20 min
                            (*ii*) 0–20% B, 30 min

SAQ 4.3a

Suggest a gradient that would improve each of the chromatograms in Fig. 4.3k. You do not need to worry about the detail, like the exact shape of the gradient, or how long it will take. Concentrate on the composition of mobile phase that is needed at the start and at the end of each chromatogram.

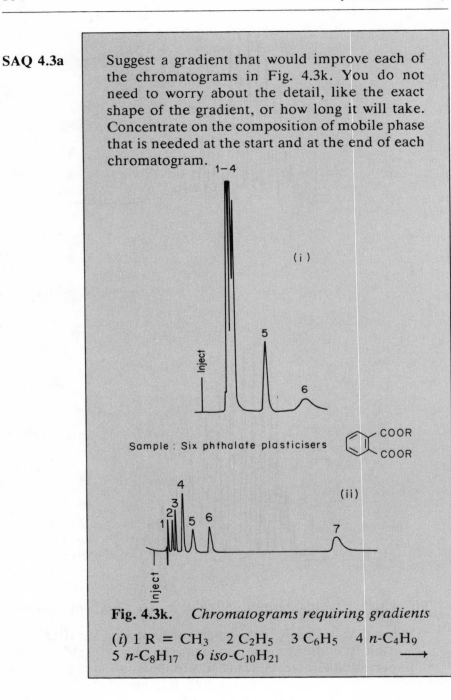

Sample : Six phthalate plasticisers

Fig. 4.3k.   *Chromatograms requiring gradients*

(*i*) 1 R = CH_3 2 C_2H_5 3 C_6H_5 4 *n*-C_4H_9
5 *n*-C_8H_{17} 6 *iso*-$C_{10}H_{21}$ $\longrightarrow$

SAQ 4.3a (cont.)

Column:	10 μm C-18 bonded phase, 30 cm × 4 mm
Mobile phase:	methanol/water 90 : 10. Flow rate 2 cm^3 min^{-1}
Detector:	Uv absorption, 254 nm

(*ii*)

Sample:	1 benzene 2 diphenyl ether 3 ethyl benzoate 4 carbazole 5 nitrobenzene 6 diphenyl ketone 7 benzyl alcohol
Column:	5 μm C-18 bonded phase. 30 cm × 1.8 mm
Mobile phase:	dichloromethane/hexane 40 : 60. Flow rate 2 cm^3 min^{-1}
Detector:	Uv absorption, 254 nm.

SAQ 4.3a

Summary

With isocratic elution and a sample having solutes with a wide range of polarity it is sometimes not possible to achieve the desired resolution in an acceptably short time. It may be possible to improve the chromatogram using gradient elution. A practical example of the development of a gradient is discussed.

Objectives

You should now be able to:

- recognise a chromatogram where the separation would be improved by the use of a gradient;

- describe a suitable gradient for a normal and a reverse phase separation;

- appreciate some of the disadvantages of using gradients.

4.4. QUANTITATIVE ANALYSIS

This example will illustrate the quantitative analysis of aspirin (acetylsalicylic acid), phenacetin and caffeine in a mixture. The structures of these three are shown in Fig. 4.4a.

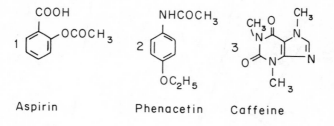

Fig. 4.4a. *Structures of aspirin, phenacetin and caffeine*

Analgesic tablets often contain aspirin and caffeine, and we will eventually use the results for the quantitative analysis of a commercial tablet.

These three are no problem to separate, they have been done on a variety of stationary phases, for instance ion exchange or reverse phase bonded silica. We will simply use a recipe, taken from the literature (G.B. Cox et.al., *J. Chromatography*, 1976, **117**, 269–78). The column is 12.5 cm × 4.6 mm with a 5 μm silica scx bonded phase. The solvent is 0.05 mol dm^{-3} ammonium formate + 10% ethanol, pH 4.8, pumped at 2 cm^3 min^{-1} with an inlet pressure of about 117 bar. Using these conditions, the compounds are separated in about three minutes (see Fig. 4.4c).

The obvious choice for detection is by uv absorbance. Fig. 4.4b shows the uv absorption spectra (from 300–225 nm) of the three compounds, made up in the mobile phase.

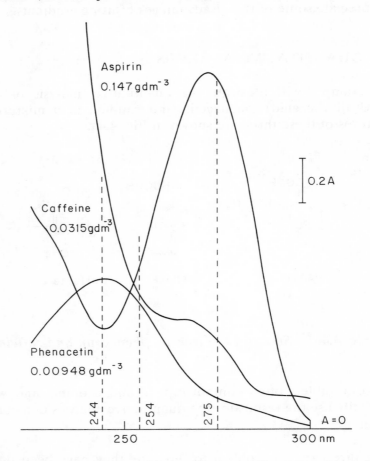

Fig. 4.4b. *Uv absorption spectra of aspirin phenacetin and caffeine*

Π   Which of the three wavelengths, 275, 254 or 244 nm would you choose for the detection of these compounds?

I think 254 nm is the least attractive of the three. At this point the absorbance of each compound is changing quite rapidly with wavelength, and we want, if possible, to monitor the absorbance at a relatively flat region in the spectrum (see Section 2.4.2). Detection at 275 nm would give a high sensitivity for caffeine, but it might represent rather a low sensitivity for phenacetin (although phenacetin absorbs much more strongly than the other two). I would choose 244 nm, which corresponds to a minimum and a maximum respectively in the spectra of caffeine and of phenacetin, although I don't think there is much to choose between 244 and 275 nm.

Fig. 4.4c shows the chromatogram obtained from a 1 μl injection of a mixture of the three compounds. The mixture was made up as follows: 0.6015 g of aspirin + 0.0765 g of phenacetin + 0.0924 g of caffeine were dissolved in 10 cm^3 of absolute ethanol. 10 cm^3 of 0.5 mol dm^{-3} ammonium formate was added, and the solution was made up to 100 cm^3.

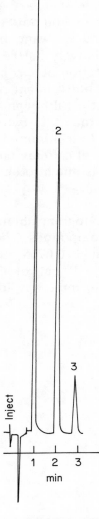

Fig. 4.4c.   *Separation of aspirin, phenacetin and caffeine*

Column: 5 μm silica scx, 12.5 cm × 4.9 mm
Mobile phase: 0.05 mol dm^{-3} H.COONH$_4$ + 10% ethanol,
                 pH 4.8.
Flow rate: 2 cm^3 min^{-1}

Detector: Uv absorption, 244 nm
Peaks: 1 = aspirin, 2 = phenacetin, 3 = caffeine

Integrator display	
Time	Area
61	144090
121	159516
170	43057

Peak areas on the chromatogram were measured with an integrator. The integrator prints out the retention time for each peak, together with a number that is proportional to the peak area.

∏   (*a*) Normalise these peak areas (this means, express each peak area as a percentage of the total peak area)

       (*b*) Tabulate the normalised areas and compare them with the known percentages of each compound in the mixture (Fig. 4.4d).

	mass in mixture, g	% in mixture	area	normalised area, %
Aspirin	0.6015			
Phenacetin	0.0756			
Caffeine	0.0924			

Fig. 4.4d. *To be completed*

The correct results for this Figure, and for the subsequent Fig.4.4e, are given at the end of the Section.

(*c*) Do the normalised areas agree with the true percentages

(*d*) What is the reason for the disagreement, if any?

None of the results agree. If you look again at the spectra in Fig. 4.4b, bearing in mind the amounts of the compounds that were used, you should be able to see the problem, which is that the detector does not show the same response for the same amount of each substance.

To allow for this, before the peak areas are normalised we must first correct each area so as to get the area we would have obtained had the detector response been the same for each of the three compounds. We will now use the results from our mixture to determine calibration factors (relative response factors) for the detector, and then use these for the analysis of a commercial tablet.

Relative response factors (r) are defined as follows:

$$r = \frac{\text{peak area of substance/mass of substance}}{\text{peak area of standard/mass of standard}} \tag{4.4a}$$

Once response factors have been obtained, using data from a standard mixture, we can then use them to correct peak areas for an unknown mixture of the same compounds, and determine the mixture by normalising the corrected areas.

Fig. 4.4e gives the peak areas obtained for three 1 μl injections of the mixture.

Π For each of the three runs shown in the Fig., taking phenacetin as the internal standard, use Eq. 4.4a to calculate response factors relative to phenacetin = 1. Average the response factors for each compound, and put the results in the Fig. below.

Injection no.		Aspirin	Phenacetin	Caffeine
	mass in mixture, g	0.6015	0.0765	0.0924
1	peak area	144090	159516	43057
	r		1	
2	peak area	143200	163164	43099
	r		1	
3	peak area	121297	139796	36564
	r		1	
	Average r		1	

Fig. 4.4e. *To be completed*

When you are sure that you have understood this method of quantitative analysis, try the following calculation:

SAQ 4.4a

A commercial analgesic tablet is stated on the packet to contain 325 mg of aspirin and 50 mg of caffeine per tablet. Two tablets + 0.0773 g of phenacetin were shaken with 10 cm^3 of ethanol for 10 minutes, then 10 cm^3 of 0.5 mol dm^{-3} ammonium formate were added and the mixture made up to 100 cm^3. The tablets contain insoluble excipients, so a little of the solution was filtered before chromatography $\longrightarrow$

SAQ 4.4a
(cont.)

Fig. 4.4f gives peak areas obtained for two 1 μl injections of the solution. Chromatographic conditions were as described in Section 4.4

Injection no.	Peak area		
	Aspirin	Phenacetin	Caffeine
1	157595	170804	50693
2	153541	164174	48478

Fig. 4.4f

Using the relative response factors that you calculated in the last Section, correct each peak area by dividing by the appropriate response factor. For each injection, calculate the amount of aspirin and caffeine present, expressing the results as mg per tablet.

The aspirin content should be between 95 and 105% and the caffeine content between 90 and 110% of the amount stated on the packet. Are the tablets within specification?

SAQ 4.4a

	mass in mixture, g	% in mixture	area	normalised area, %
Aspirin	0.6015	78.2	144090	41.6
Phenacetin	0.0756	9.8	159516	46.0
Caffeine	0.0924	12.0	43057	12.4
		100.0		100.0

Completed Fig. 4.4d

Injection		Aspirin	Phenacetin	Caffeine
	mass in mixture, g	0.6015	0.0756	0.0924
1	peak area	144090	159516	43057
	r	0.1135	1	0.2208
2	peak area	143200	163164	43099
	r	0.1103	1	0.2160
3	peak area	121297	139796	36564
	r	0.1090	1	0.2139
	Average r	0.111	1	0.217

Completed Fig. 4.4e

Summary

From the uv absorption spectra, a suitable wavelength is found for the simultaneous detection of aspirin, phenacetin and caffeine. Using phenacetin as internal standard, response factors are calculated for aspirin and caffeine and the results are used for the quantitative determination of aspirin and caffeine in an analgesic tablet.

Objectives

You should now be able to:

● use uv absorption spectra to identify a suitable detection wavelength to use for the quantitative analysis of a mixture;

● appreciate that for quantitative analysis the detector has to be calibrated;

- calculate relative response factors from data obtained with standard mixtures, and use the response factors for the quantitative determination of an unknown.

5. Some Practical Aspects of hplc

5.1. PACKING COLUMNS

Microparticulate stationary phases are packed into columns by forcing a slurry of the packing material in a suitable solvent into the column under high pressure. Manufacturers of columns are sometimes secretive about the methods they use for this, thus lending support to the view that the successful packing of hplc columns is as much of an art as a science. If you need very high efficiency, or good reproducibility between columns, it is better to use manufactured columns. If not, you can make your own columns using simple and fairly inexpensive equipment. A number of suppliers sell systems for column packing, but it is much cheaper to buy the items individually and assemble them yourself.

Although there is no standard method for packing hplc columns, there is a general consensus about the experimental conditions that are required. These are:

(*a*) The stationary phase particles must be properly dispersed in the slurry and must not coagulate;

(*b*) Sedimentation of the stationary phase should be avoided during the packing process;

(*c*) The stationary phase particles should hit the column bed with a high impact velocity;

(*d*) The bed should be packed under high compression.

Preliminary dispersion of the stationary phase in a suitable solvent is best carried out in an ultrasonic bath. To prevent sedimentation of the stationary phase during packing, a number of different approaches have been used:

(*a*) The stationary phase is dispersed in a solvent mixture which is formulated so as to have to same density as that of silica (2.2 g cm^{-3}). As one of the solvents must have a density greater than 2.2, the choice is rather restricted. For example, 1,1,2,2-tetrabromoethane (density 2.86) and tetrachloroethene (density 1.62) could be used.

∏   What composition would give the right density, assuming that the density of the mixture varies linearly with composition?

If v = volume fraction of $C_2H_2Br_4$, then

$$2.2 = 2.86v + 1.62(1 - v)$$

Hence, $C_2H_2Br_4 = 47\%$, and $C_2Cl_4 = 53\%$

Because the halogenated hydrocarbons that have to be used for this are both toxic and expensive, the use of balanced density slurries for packing columns is declining.

(*b*) Another method of reducing sedimentation in the packing slurry is to use a viscous solvent (eg glycerol/methanol mixtures).

∏   What do you think would be the difficulty with this method?

To achieve a reasonable flow rate through the column during the packing process (condition (*c*) above), very high pressures have to be used. Pressures greater than 1700 bar (25000 psi) have been used for packing columns using viscous solvents. The apparatus has to be designed to withstand these high pressures, and consequently becomes expensive.

(*c*) A third approach is to use low density low viscosity solvents like methanol or propanone. With these, a satisfactory flow rate through the column during packing (about 15 cm^3 min^{-1} for 5 μm silica) can be obtained with relatively low pressures (roughly 350–650 bar, or 5000–10000 psi). Sedimentation is minimised by not wasting time in those parts of the packing procedure where sedimentation can occur. In some methods, during packing, the slurry is contained in a reservoir fitted with a stirrer.

Fig. 5.1a shows the slurry packing system that I use. This operates at a fairly low pressure, as I am not especially interested in producing columns with very high efficiencies, but rather in saving on the cost of commercial columns. The pump and high pressure valve are rated for pressures of 500 bar (7500 psi) and 400 bar (6000 psi) respectively. The slurry reservoir is a stainless steel tube 85 cm long, with a capacity of about 50 cm^3. The method is not hazardous unless there is air trapped in the high pressure line; nevertheless it is advisable to use a safety screen.

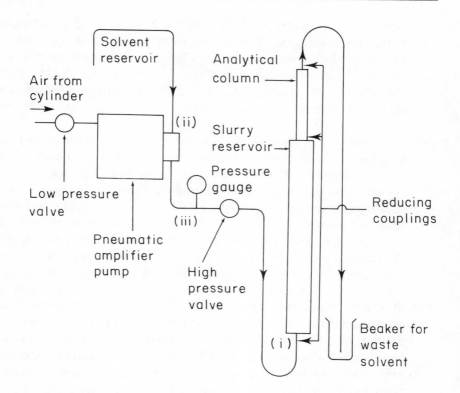

Fig. 5.1a. *Column packing system*

This is the method I use for packing a column with 5 μm bonded phase silica:

(*a*) The column tubing is washed with tetrachloromethane, then propanone, and finally dried. It is then fitted with a 1/4 inch cap at the end that is to be attached to the slurry reservoir, and a 1/4–1/16 inch zdv reducing coupling, fitted with a 1 μm stainless steel gauze, at the other end,

(*b*) For a 12.5 cm column, approximately 1.7 g of the silica is added to 30 cm^3 of methanol and the slurry is stirred with a magnetic follower until just before it is placed in the slurry reservoir. This quantity of silica is slightly more than is needed to fill the column, (*c*) The packing solvent (400 cm^3 of methanol + 0.2 g of sodium ethanoate) is degassed under vacuum for 10 minutes and then placed in the solvent reservoir. The sodium ethanoate is added to prevent static build up on the stationary phase during packing, which can produce unstable column beds, especially with bonded phase packings,

(*d*) With the low pressure valve closed, the air pressure is adjusted to about 6.7 bar (100 psi) using the second stage regulator on the air cylinder. The slurry reservoir is disconnected at point (*i*) and the pump is tested by opening the low pressure valve slightly and then opening the high pressure valve. If there is air trapped anywhere in the line the pump may not work at all, or the flow of solvent from point (*i*) may not be fast enough (this decision requires a bit of experience). If the solvent flow appears satisfactory, the high pressure valve is closed, when the pressure on the Bourdon gauge should rapidly increase (to the value set by the inlet pressure and the amplification of the pump). If the pressure increases only slowly, this indicates the presence of air in the line. Air pockets are often present if the system has not been used for some time; if this is the case, I usually do step (*d*) first. To remove trapped air, the solvent line is disconnected at points (*ii*) and (*iii*) and solvent is passed rapidly through each stage, using a 20 cm^3 syringe with a 1/16 fitting attached to the needle,

(*e*) Assuming the pump is behaving properly, the solvent line is connected at point (*i*) the high pressure valve is closed and the system is pressurised to 350 bar (about 5000 psi) by opening the low pressure valve. The remaining operations in (*e*) are carried out as quickly as possible, to limit the effects of sedimentation. The slurry is poured into the reservoir, which is then topped up with methanol. The column and outlet tube are connected and then the high pressure valve is opened, (*f*) After about 200 cm^3 of solvent have been pumped through the column (about 15 minutes), the column and slurry reservoir are inverted, pumping is continued for a further 5 minutes, then the high pressure valve is closed and the pressure on the pump side released by closing the low pressure valve. After 10 minutes the column is disconnected from the reservoir, the top of the packing is smoothed with a razor blade and a gauze and reducing coupling are fitted,

(*g*) The column is then attached to the injection unit on the chromatograph (but not to the detector) and mobile phase is pumped through the column for 10 minutes at about 3 cm^3 min^{-1}. The column is then connected to the detector and pumping is continued at a lower flow rate until (with a uv absorbance detector) a stable baseline can be obtained on the lowest absorbance setting.

When the column is ready to be used, the chromatogram of a suitable test mixture should be obtained. The plate number and retention times of the test solutes should be noted, and the peaks should have a satisfactory shape (minimal tailing). For measurement of the plate number, the recorder should be used at a high chart speed. Fig. 5.1b(*i*) and (*ii*) show test chromatograms for a C-18 column prepared by the above method, and Fig. 5.1c and 5.1d show the data that you should report with the chromatogram. The retention for an unretained peak is taken as the small baseline disturbance just before the first peak.

∏   See if you can complete the data in Fig. 5.1d from measurements on the chromatogram run at the higher chart speed. For reasons of space, the chart is marked at 10 cm past the injection point, so measure retention distances from the mark and add 10 cm. The correct values are given at the end of the Section.

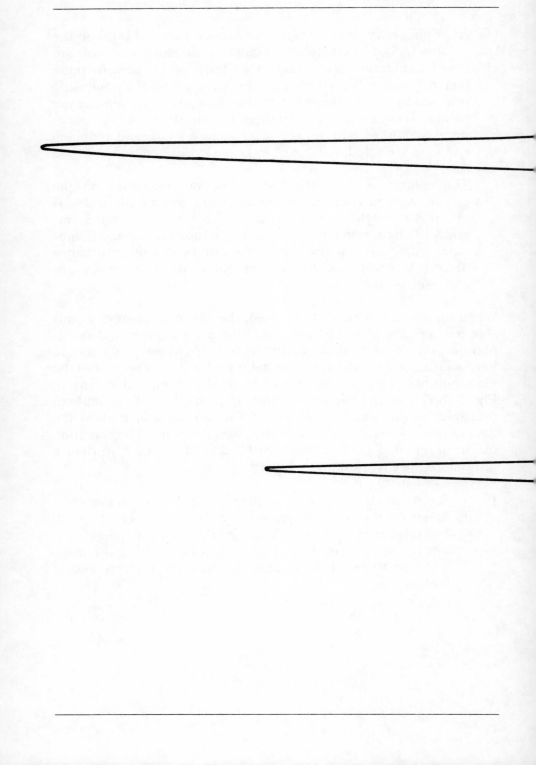

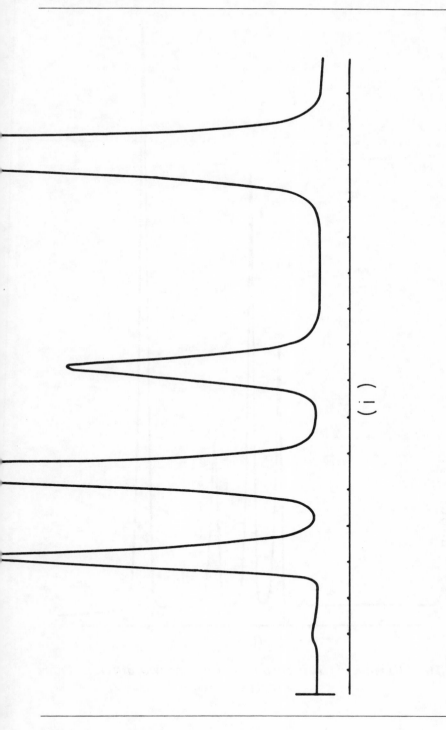

(i)

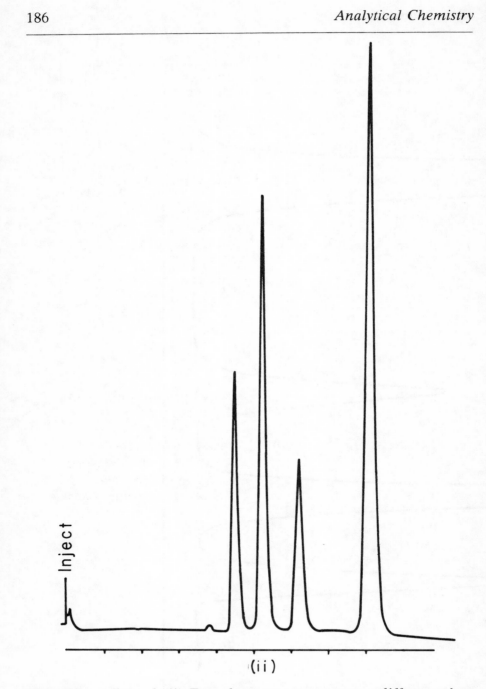

Fig. 5.1b.   *(i) and (ii) Test chromatograms at two different chart speeds*

Column length	12.5 cm
Internal diameter	4.9 mm
Stationary phase	5 μm C-18
Mobile phase	CH_3OH/H_2O 60:40
Flow rate (nominal)	1 cm^3 min^{-1}
Pressure drop	83 bar (1250 psi)
Temperature	ambient (about 22 °C)
Detector	uv absorption, 254 nm, 0.5 aufs
Injection volume	0.5 μl
Recorder	10 mV, 10 mm min^{-1} and 1 mm s^{-1}
Test mixture (peaks 1–4)	propanone, phenol, 4-hydroxymethylbenzene, methyl phenyl ether, dissolved in mobile phase.
Unretained solute retention distance	116 mm

Fig. 5.1c. *Conditions used for chromatograms in Fig. 5.1b*

Peak	1	2	3	4
Retention distance, mm		162		252.5
$w_{\frac{1}{2}}$, mm		6		8
k'		0.39		1.18
N (Eq. 2.6c)		4038		5519

Fig. 5.1d.   *To be completed*

The column was satisfactory apart from the pressure drop, which was high, indicating a partial blockage, probably of one of the gauzes.

In order to measure k' for the solutes we have to know the retention distance or volume of an unretained solute. The accurate determination of this quantity is not an easy problem with bonded phase columns and reverse phase operation. A number of different methods have been used:

(a) Determination of the difference in mass between the column full of solvent and the dry column;

(b) Injection of a homologue of one of the mobile phase components, with detection by refractive index. For example, with a methanol/water mobile phase, ethanol could be used;

(c) Injection of D_2O or labelled components of the mobile phase;

(d) Mathematical methods based on retention data obtained for a homologous series of compounds;

(e) Injection of unretained solutes, eg $NaNO_3$ or tartrazine, which can be detected by uv absorption. The idea is that such highly polar solutes will not be retained by the nonpolar stationary phase. Sometimes, the flow disturbance before the first peak is used (as above) or even the first peak itself is taken as an unretained solute, if it is a very polar compound.

The trouble is that the various methods all give different results. In particular, the gravimetric method usually gives higher results than the others. The gravimetric method will measure the total volume of solvent in the column, ie void volume + interstitial volume. Lower results will be obtained from the injection of solutes if the solutes are partly excluded from the pores of the stationary phase. If the measured retention distance from the chromatogram is converted into a volume, the flow rate from the pump must be carefully determined by collecting the mobile phase for a known time and weighing it.

Peak	1	2	3	4
Retention distance, mm	138	162	190	252.5
$w_{\frac{1}{2}}$, mm	5.5	6	7	8
k'	0.19	0.39	0.64	1.18
N	3488	4038	4081	5519

Completed Fig. 5.1d

Summary

Techniques for packing hplc columns and for testing the packed column are described.

Objectives

You should now be able to:

● appreciate the experimental conditions needed for the successful packing of hplc columns;

● describe how a column is packed;

● understand how to evaluate a test chromatogram.

5.2.   THE PREPARATION OF MOBILE PHASES

This section describes some of the problems that can occur with the mobile phase in hplc. Many of these problems arise because of the presence of impurities, additives, dust or other particulate material, or dissolved air. It is always best to try to prevent these problems by a little attention to detail and the use of simple good housekeeping procedures. Although it is always tempting to try to save time and expense by the neglect of such matters, if you do this you will store up trouble for yourself in the long term.

A major cause of practical problems in hplc is the presence of air bubbles in the mobile phase at some point in the system. Some of the symptoms of trapped air were dealt with in Section 2. Air bubbles can collect in the pump, or the detector cell, or in other places. Because of their compressibility, air bubbles will reduce the volume of mobile phase delivered by the pump so that reproducibility is affected, and because of the flow variation, the detector noise is often worse as well. Large air bubbles in the pump may stop the pump working. Detection can be affected in various ways. For example, with uv absorbance detectors, air in the detector cell can cause serious noise, or very high absorbance. Dissolved oxygen can interfere with uv absorbance detection at short wavelengths, as oxygen absorbs radiation strongly below 200 nm. Many problems with dissolved air are avoided if the mobile phase is degassed before use, so this should always be done. Degassing can be accomplished by placing the mobile phase under vacuum, or by heating and ultrasonic stirring, or by a combination of these. In many practical arrangements, having removed the air, the mobile phase is then placed in a reservoir in contact with air, allowing the uptake of air to start again. If this is the case, degassing should be repeated every hour or so. To restrict access of air to the mobile phase, some arrangements use a straight sided reservoir with a plastic float that sits on top of the mobile phase. Another technique involves saturating the mobile phase with helium, which has a small solubility in liquids. Access of air is restricted by having a continuous slow stream of helium passing over the mobile phase in the reservoir. Although helium is rather expensive, only a small amount is used, and this method is now very popular.

A microparticulate hplc column is a very efficient filter, and if the mobile phase contains any particulate matter, or acquires it from the pump and/or the injection valve due to wear, it will collect at the top of the column. If this happens, the pressure drop across the column for a given flow will gradually increase, and the column may eventually become completely blocked. To prevent this happening, the mobile phase should always be filtered before use, preferably through a 0.5 μm porosity filter, and guard and scavenger columns should be used as a matter of routine (see Section 5.3.2).

Many reagent grade solvents contain levels of impurities that make them unsuitable for long term use in hplc. Sometimes the impurities are added deliberately as antioxidants, stabilisers, or for denaturing. Wherever possible, 'hplc grade' solvents should be used to prepare mobile phases, or alternatively the solvents should be adequately purified before use.

Distilled or deionised water contains small amounts of organic impurities which can cause problems in long term use with bonded phase columns in the reverse phase mode. The non-polar stationary phase will collect these organics, which can alter the nature of the stationary phase or sometimes produce spurious peaks (Fig. 4.3c is an example of this). Water purification can be done by distillation from permanganate, by passage of the water through bonded phase columns, or by means of commercial systems, eg the Millipore system which uses a combination of carbon filters, a mixed bed ion-exchanger and a 0.2 μm filter at the outlet. This removes traces of inorganic, organic, and particulate material from the water.

Impurities in other solvents may affect chromatographic behaviour, or detection, or both. Chlorinated solvents such as di- or trichloromethane are stabilised against oxidative breakdown by the addition of small amounts of methanol or ethanol.

∏   In a normal phase separation using a trichloromethane/ heptane mobile phase, how would the presence of stabiliser in the trichloromethane affect the separation?

The presence of alcohol would increase the polarity of the mobile phase, so that solute retention times would be shortened. Also, we would not expect to get very good reproducibility, as the concentration of stabiliser will vary slightly from batch to batch.

Chlorinated hydrocarbons can be purchased without stabiliser, or the stabiliser can be removed by adsorption onto alumina, or by extraction with water, followed by drying. Unstabilised di- or trichloromethane will slowly decompose, producing HCl, which corrodes stainless steel. The rate of decomposition may be accelerated by the presence of other solvents.

Ethers contain additives to stabilise them against peroxide formation. For instance, tetrahydrofuran is commonly stabilised by the addition of small amounts of hydroquinone. This absorbs uv radiation strongly and so interferes with uv absorbance detection. It can be removed by distilling the solvent from KOH pellets. If you use inhibitor-free tetrahydrofuran, it should be stored in a dark bottle and flushed with nitrogen after each use. Any peroxides that form should be periodically removed by adsorption onto alumina.

When mixing solvents to form mobile phases, the volume of each component should be measured separately before the solvents are mixed, since the volume of the mixture does not usually equal the sum of the two separate volumes. For example, 50 cm^3 of methanol mixed with 50 cm^3 of water produces a total volume of about 96 cm^3 of a 1:1 mixture. If you make up a mobile phase by filling a measuring cylinder half full of methanol and then making up to the mark with water, you won't end up with a 1:1 mixture. If the mobile phase contains volatile components, the composition can alter during degassing procedures. Measurement of the refractive index of the mobile phase is a useful check on the composition.

With uv absorbance detectors, we have to consider the uv absorption of the mobile phase, which always increases as the wavelength decreases. The 'uv cut-off' of solvents indicates the useful wavelength range of the solvent and means the wavelength below which the solvent has an absorbance of 1 or more when measured in a 1 cm cell. Aliphatic hydrocarbons cut off at about 210 nm; the best polar solvents for low wavelength work are methanol and acetonitrile, which cut off at 205 and 190 nm, respectively, provided they are pure. Acetonitrile is difficult to purify, and is consequently expensive.

Other properties of hplc solvents that we may need to consider include compressibility, viscosity, refractive index, vapour pressure, flash point, odour and toxicity. Most hplc textbooks contain tables of these properties. For instance, there is a useful table in the book edited by J. H. Knox.

The solvent properties listed in Fig. 5.2a are used in SAQ 5.2b.

	viscosity cp, 20 °C	boiling point, °C	uv cut-off, nm	price	TLV*
pentane	0.23	36.2	210	10.60(*i*)	500
heptane	0.43	98.4	210	8.70	500
trichloromethane	0.57	61.2	245	5.30	50
tetrachloromethane	0.97	76.8	265	5.80(*i*)	10
acetonitrile	0.37	82.0	190	6.80	40
dioxane	1.54	101.3	220	10.10	100
methanol	0.60	64.7	205	3.40	200
ethanol	1.20	78.5	210	19.20(*ii*)	1000
propan-2-ol	2.30	82.3	210	5.90	200
propanone	0.56	56.5	330	5.00	1000

Fig. 5.2a.   *Properties of solvents commonly used in hplc*

Prices are given in £/dm^3 (1986) for BDH HiPerSolv hplc solvents, except for:

(*i*) AnalaR reagent; (*ii*) 99.7–100% AnalaR reagent, less duty

* Threshold limit values (TLV) are in ppm by volume at 25 °C and 760 mm and are taken from the *Handbook of Chemistry and Physics*, 54th Edn., 1973–74.

SAQ 5.2a

The following account, taken from a practical notebook, describes the preparation of the mobile phase used in the example in Section 4.4.

$\longrightarrow$

**SAQ 5.2a
(cont.)**

'1.575 g of ammonium formate was made up to 500 cm^3 in a graduated flask. To this solution, 50 cm^3 of ethanol was added, and after mixing the mobile phase was placed in the solvent reservoir and pumping was commenced at 2 cm^3 min^{-1}'.

Can you identify three mistakes that were made?

SAQ 5.2b   These questions refer to the properties of solvents that were listed in Fig. 5.2a.

(*i*)   Considering a heptane/trichloromethane mobile phase, why would it be undesirable to replace the heptane by pentane, or the trichloromethane by tetrachloromethane?

(*ii*) Why are methanol or acetonitrile preferred to other water miscible solvents for the preparation of mobile phases for reverse phase chromatography?

(*iii*) What would be the major difficulty associated with the use of propanone in mobile phases?

SAQ 5.2b

Summary

Problems can occur with the mobile phase because of the presence of particulate matter, impurities or dissolved air. Some of the practical remedies are considered.

Objectives

You should now be able to:

● appreciate the need to use mobile phases of sufficient purity;

● recognise some of the symptoms of the presence of air or other impurities in the mobile phase;

● identify the methods used to degas mobile phases.

5.3. PRACTICAL PROCEDURES WITH COLUMNS AND SAMPLES

5.3.1. Care of hplc Columns

The useful lifetime of hplc columns is shortened by the appearance in the column packing of cracks or voids, especially at the top of the column, or by the collection at the top of the column of particulate material from the mobile phase (eg wear particles from the pump or injection valve), or strongly held components from the sample.

The appearance of voids at the top of the column bed is a common problem, which can be caused by the gradual settling of the column packing or by the dissolution of the silica stationary phase, caused, for instance, by the use of alkaline mobile phases. The presence of a void at the top of the column will result in a loss of efficiency because of the extra dead volume in the system. If the profile of the column bed at the top of the column is irregular, a single solute may produce a peak with a shoulder, a double peak or even a group of peaks. What happens is that as the solute enters the column it encounters the stationary phase at different points in time, and the end result is as though two or more chromatograms are superimposed, all slightly out of phase. The effect is shown in Fig. 5.3a; Fig. 5.3b shows a practical example of this type of behaviour.

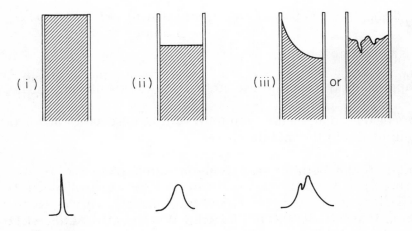

Fig. 5.3a.   *Packing faults at the top of the column*

A void at the top of the column, (ii), produces additional dispersion. Uneven packing as in (iii) adds to the dispersion and may cause a solute to appear as two or more peaks.

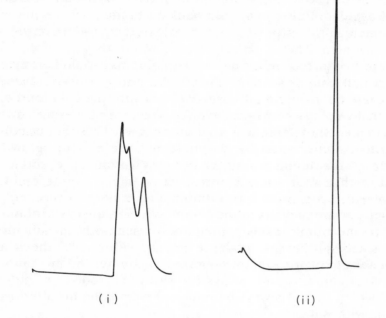

(i)   (ii)

Fig. 5.3b. *(i) Chromatogram obtained with defective packing at the top of the column*

Column: 5 μm C-18 12.5 cm × 4.9 mm
Mobile phase: CH_3OH/H_2O 60 : 40. Flow rate: 1 cm^3 min^{-1}
Detector: Uv absorption, 254 nm Sample: propanone

(ii) The same sample after repair of the top of the column.

The packing at the top of the column can be repaired by first re-moving the inlet compression fitting and frit. A small quantity of the packing is then scraped out with a small screwdriver or micro-spatula, then the packing is tamped down gently with a glass or plastic rod of the right diameter. The void can be filled with bal-lotini (glass microspheres) or a thick slurry of the stationary phase in a suitable solvent. The packing is levelled at the top with a razor blade, and then the frit and compression fitting are replaced.

Although hplc columns are packed under high pressure, the column bed can be disturbed by sudden pressure variations, or mechanical or thermal shock, so to extend the lifetime of columns all of these should be avoided. Columns should be taken gradually from low to high or from high to low pressures and sudden temperature changes should be avoided. They should be stored where they are unlikely to be bumped or jarred. When not in use, columns should be stored in an inert non-volatile solvent. Usually the storage solvent should have the same polarity as that normally used with the column. For instance, bonded phase columns for use in reverse phase mode can be stored in methanol or methanol/water mixtures. These are usually preferred to distilled water, as the methanol prevents the growth of bacteria. The columns should be securely capped at each end, preferably with compression fittings. Always make a note of the storage solvent that is used; if you do not, you may have miscibility problems the next time you use the column. For instance if a column is stored in methanol and is subsequently used with an aqueous buffer, it is necessary to first wash the column with water, otherwise the buffer salts may precipitate when they encounter the methanol. Stainless steel columns and fittings are slowly corroded by buffer solutions, which should always be removed from columns after use by washing with water.

5.3.2.   Protection of Columns During Use

The useful life of an analytical column is subsequently improved by the use of pre-columns, located between the pump and injector (this sort are sometimes called *scavenger* columns) and/or between the injector and the analytical column (*guard* columns). The scavenger column is an efficient final filter that protects the analytical column from wear particles from the pump and from dust or other particulate matter in the mobile phase. Guard columns protect the analytical column from wear particles from the injection valve and from any adverse characteristics of the sample. Complex samples may contain components that are irreversibly held on the stationary phase under the conditions of the separation. If these materials build up at the top of the column they can have a profound effect on chromatographic behaviour such as retention, selectivity and efficiency. What we are doing with guard and scavenger columns is to transfer these problems from a relatively expensive analytical column to a relatively cheap pre-column.

Because scavenger columns are located upstream of the injection valve, they do not add to the dispersion of the chromatogram, and their size is not critical in this respect. Guard columns, on the other hand, do cause a slight loss of efficiency, and so need to have a relatively small volume. Reducing the volume, of course, reduces the life of the guard column.

For use with 25 cm × 4.6 mm analytical columns, guard columns and scavenger columns are often 4.6 mm internal diameter and 3–10 cm in length. They can be packed with microparticulate stationary phases or with porous layer beads. Porous layer beads are cheaper than microparticulates and are easier to pack, but they have lower capacities and will require changing more often. It is usually difficult to know how long a pre-column will last before it requires changing. In routine work, precolumns are usually repacked or replaced to a fixed schedule.

An alternative use of pre-columns is in sample pre-concentration steps. Solutes present at very low levels in a sample can sometimes be concentrated by passing a large volume of sample through a small column on which the solutes are strongly retained. This column can then be connected to the analytical column and the solutes eluted with a suitable mobile phase. The pre-column can be packed with a stationary phase having a relatively large particle size so that the sample can be pumped through rapidly at low pressure, using a cheap pump. Pre-columns filled with high-porosity silica have also been used in work with basic mobile phases to extend the lifetime of the analytical column by saturating the mobile phase with silica.

If the performance of a column is no longer satisfactory it can sometimes be reconditioned by washing with a suitable solvent, or series of solvents. Some bonded phase columns, C-18 for instance, tend to collect non-polar impurities, which can sometimes be removed by washing the column with a non-polar solvent, eg heptane. Assuming the mobile phase normally used with the column is CH_3OH/H_2O 50:50, we cannot wash directly with heptane; because of miscibility problems, we have to get to heptane *via* a miscible solvent or series of solvents.

∏   How could this be done?

You could first wash the column with methanol, then trichloromethane, then heptane (or methanol, ethyl ethanoate, heptane). You cannot go directly from methanol to heptane because the two are only partly miscible. The column needs to be washed with about 20 dead volumes of each solvent (about 50 cm^3 of each solvent for a 25 cm × 4.6 mm column). To get back to CH_3OH/H_2O 50:50 you would have to go through the sequence of solvents in reverse. If buffer solutions or ion-pairing reagents have been used in the mobile phase, very much longer equilibration times may be needed.

5.3.3.   Sample Preparation and Clean-up

Samples in hplc can come from a very wide range of sources, and unfortunately they are usually not simple mixtures of pure organic compounds, all of which are soluble in and separable by the same mobile phase. Biological samples may contain proteins, salts and a host of organic compounds with widely differing polarities. Pharmaceutical samples often contain a wide range of soluble and insoluble excipients. Samples to be analysed for environmental pollution can and do contain almost anything! Very often, we are concerned with the separation of sample components of interest, sometimes present at very low levels, from a range of other components in the sample matrix which may interfere with the analysis.

Many of the adverse consequences of injecting dirty samples can be prevented or minimised by the use of guard columns, as discussed earlier, but often some form of sample clean-up is needed as well. The goal of sample preparation is to obtain, from the sample, the components of interest in solution in a suitable solvent, free from interfering constituents of the matrix, at a suitable concentration for detection and measurement. Naturally we want to do this with the minimum time and expense.

Whenever possible, the sample should be dissolved in the same solvent or mixture that is used for the mobile phase. If a different solvent is used, this often results in loss of efficiency and poor peak shape. What happens is that as the injected material moves down the column, sample molecules at the edges of the injected band are in contact with a mobile phase that has a different composition to that seen by the bulk of the sample. The result is that the molecules at the band edges will travel at a different speed to that of the rest of the sample, resulting in spreading or splitting of the peaks.

Fig. 5.3c shows the consequences of dissolving a sample in solvents that are not the same as the mobile phase.

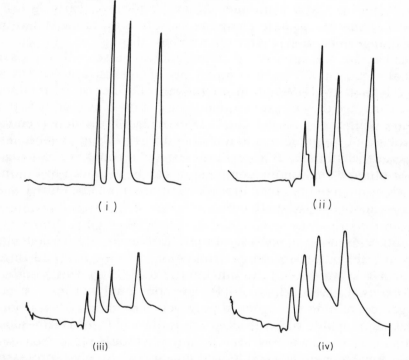

Fig. 5.3c. *Injection of a test mixture dissolved in different solvents*

Column:	5 μm C-18 bonded phase 12.5 cm × 4.9 mm
Mobile phase:	CH_2OH/H_2O 60 : 40 Flow rate: 1 cm^3 min^{-1}
Detector:	Uv absorption, 254 nm
Sample:	as in Fig. 5.1b and 5.1c, dissolved in (i) mobile phase (ii) tetrahydrofuran (iii) ethanol (iv) butanol

Traditional methods of sample preparation such as liquid–liquid or liquid–solid extraction are time consuming and can sometimes result in incomplete sample recovery. They have to a large extent been replaced by column extraction procedures. In many ways, these resemble the use of guard columns, except that the extraction or purification is done before the chromatography. Column extraction can be done conveniently using the Waters '*Sep-Pak*' cartridges, which are radially compressed plastic 'mini-columns', 2.5 cm long × 1 cm diameter, through which samples can be passed using a syringe. They are available packed with silica, alumina, or a C-18 bonded silica. With these, you can either choose conditions (cartridge and solvent) so that the sample components of interest are retained on the packing whilst matrix interferences pass through unretained, or you can retain the interfering compounds. The second approach is usually chosen when the components of interest are present at relatively high levels. When the components of interest are retained on the cartridge they can subsequently be removed by eluting with a solvent of different polarity. The C-18 cartridges are widely used for the isolation of organic compounds (eg drugs or drug metabolites) from biological fluids. After the fluid is passed through the cartridge, salts or other water-soluble components can be removed by eluting with water, then the retained organic compounds can be eluted with a less polar solvent, eg methanol.

As another example of the use of these cartridges we will consider a recent publication on the determination of some two and three ring azines in water (T.R. Steinheimer and R.G. Ondrus, *Analytical Chemistry*, 1986, **58**, 1839–44). The compounds of interest are nitrogen containing analogues of polycyclic aromatic hydrocarbons (PAHs) and include such substances as quinoline (1-benzazine) and acridine (2,3,5,6-dibenzopyridine). They are thought to be released into the environment as a result of combustion or pyrolysis processes involving fossil fuels. Some of them are highly carcinogenic. In contaminated water samples they are usually accompanied by PAHs, which are present at much higher levels. The azines in contaminated water samples were first enriched and separated from accompanying PAHs using an extraction technique on a non-polar cartridge. Details of this, and of the conditions used in the chromatography, are summarised below. After reading the summary, see if you can answer the problems in SAQ 5.3a.

A sample of water with a volume of up to 2 dm^3 was filtered through Whatman no. 1 filter paper to prevent clogging of the extraction cartridge. The filtered water sample was then passed through a Waters C-18 Sep-pak at flow rates of up to 200 cm^3 min^{-1}. The cartridge was centrifuged to remove the residual water and was then eluted with 2 cm^3 of 25 : 75 acetonitrile/0.78 mol dm^{-3} HCl. The eluate was neutralised with aqueous ammonia and then filtered through a 0.2 μm porosity nylon filter.

For the chromatographic separation of the azines, the column used was 10 cm × 8 mm Nova-pak (4 μm C-18 bonded phase). The mobile phase was acetonitrile/water 42 : 58, pH 7.2, pumped at 1.5 cm^3 min^{-1}.

SAQ 5.3a

These questions refer to the extraction and separation of azines, that was discussed in the previous Section:

(*i*) What difference between the azines and the PAHs is exploited to achieve their separation on the C-18 cartridge?

(*ii*) Why is HCl used when the cartridge is eluted?

(*iii*) If the PAHs had not been removed from the sample, how would you expect their retention times to compare with those of the corresponding azines (eg anthracene and acridine or naphthalene and quinoline)?

(*iv*) The chromatographic separation was done at pH 7.2. Can you think of any disadvantages of using a lower or a higher pH?

(*v*) Which detector could be used for the separation?

SAQ 5.3a

5.3.4. Column Switching

The use of extraction cartridges is one example of a technique known as column switching (it is also often called *multidimensional* chromatography). The method can be used, either on-line or off-line, for sample clean-up by selecting part of a complex chromatogram (a *cut*) and transferring the cut to one or more secondary columns for further separation. Alternatively column switching can be used for on-column concentration, in which a large volume of sample is passed through a pre-column under conditions such that the solutes of interest are retained. After concentration in this way, the mobile phase is changed so that the solutes are eluted rapidly, and another column is used for the analysis.

The use of extraction cartridges in the separation of azines, discussed in the last Section, is an example of on-column concentration using off-line column switching. A chromatogram can be cut off-line by collecting the zones of interest at the detector outlet followed by reinjection of the collected fraction onto a secondary column. The mobile phases used with the two columns should be compatible, eg they should be miscible and the mobile phase used with the first column should not have too high an eluting power in the second column. If the mobile phases are incompatible it may be possible to evaporate the primary mobile phase and redissolve the sample in a suitable solvent.

The example shown in Fig. 3.4g uses off-line column switching to combine exclusion and reverse phase chromatography for the separation of pesticides from a complex sample matrix.

With on-line techniques, the column switching operations are done using valves. Fig. 5.3d shows a simple arrangement for zone cutting that could be used for sample clean-up. The zone marked Y is to be determined and all other zones are to go to waste (this type of cut is called a *heart cut*). Initial separation takes place on column C1 so that early zones (X) are routed to waste. When zone Y is eluted from C1, valve V2 is switched to elute this zone onto column C2. After complete transfer of Y onto C2, valve V1 is switched to prevent further elution of unwanted zones (Z, for instance). Zone Y is eluted to the detector and C1 can be cleaned and re-equilibrated with mobile phase.

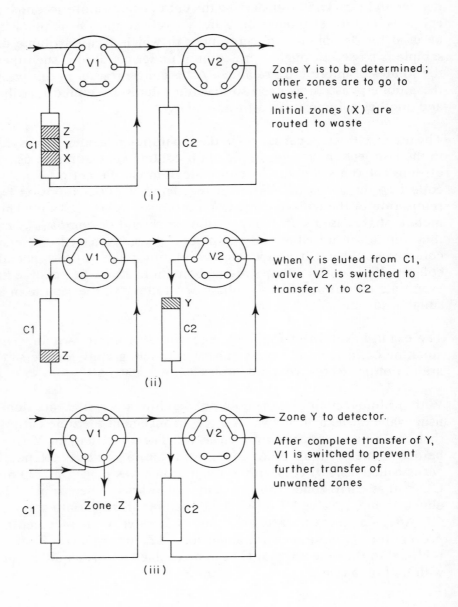

Zone Y is to be determined; other zones are to go to waste.
Initial zones (X) are routed to waste

(i)

When Y is eluted from C1, valve V2 is switched to transfer Y to C2

(ii)

Zone Y to detector.

After complete transfer of Y, V1 is switched to prevent further transfer of unwanted zones

(iii)

Fig. 5.3d. *On-line column switching*

INSERT Zone Y is to be determined; other zones are to go to waste. Initial zones (X) are routed to waste.

INSERT When Y is eluted from C1, value V2 is switched to transfer Y to C2.

INSERT Zone Y to detector.

After complete transfer of Y, V1 is switched to prevent further transfer of unwanted zones.

SAQ 5.3b Comparing on-line and off-line methods of column switching, which of the following advantages and limitations do you think would apply to the on-line technique?

(*i*) Easier to carry out

(*ii*) Cheaper

(*iii*) Faster

(*iv*) Better reproducibility

(*v*) More chance of sample loss

(*vi*) Easier to automate.

SAQ 5.3b

Summary

The useful life of hplc columns can be extended by proper treatment, in particular by the use of guard and scavenger columns. Pretreatment of samples is often necessary, eg very dilute samples may require concentration, or complex samples may need to be cleaned up. Some of the more important techniques are considered.

Objectives

You should now be able to:

● appreciate some of the conditions needed to prolong the life of hplc columns;

● understand the function of guard columns and scavenger columns in protecting the analytical column;

● recognise that column switching techniques can be useful in the treatment of dilute or complex samples.

Self Assessment
Questions and Responses

SAQ 1a

Complete the following definition of liquid chromatography by filling in the blanks. For each space, choose a word from the groups given below:

Liquid chromatography is a technique for the .... of mixtures in which the sample is introduced into a system of two .... Differences in .... shown by the solutes cause them to travel at different speeds in the ....

(*i*)   analysis   (*iii*)   adsorption
     separation   distribution
     determination   partition

(*ii*)   substances   (*iv*)   liquid
     chemicals   mobile phase
     phases   system

Response

(*i*)   'Analysis' or 'determination' are not too bad, but the real power of chromatography is as a separation method, so 'separation' is the one I would choose.

(*ii*)   'Phases' indicates the principle of the technique, the other two are too vague.

(*iii*) 'Distribution' is best (the other two are different kinds of sorption mechanism and are too specific).

(*iv*)   'Liquid' is not correct, as both the mobile and stationary phase may be liquids. 'Mobile phase' is not right either, as the solutes all travel at the same speed when they are in the mobile phase. So you are left with 'system', meaning, 'combination of stationary and mobile phases'.

****

SAQ 1b   For which of the following would hplc be a suitable means of analysis?

(*i*) Determination of the composition of cigarette lighter fuel.

(*ii*) Analysis of ascorbic acid (vitamin C) in a vitamin C tablet.

(*iii*) Determination of the amount of caffeine in a soft drink.

(*iv*) Separation of a mixture of naturally occurring sugars.

(*v*) Separation of a mixture of amines.

Response

(*i*) This will be a mixture of light hydrocarbons and would be a clear case for gc, which would give an easier, quicker and cheaper separation than hplc.

(*ii*) The tablets will contain ascorbic acid, together with insoluble fillers and binders. The only separation needed would be to take up the tablets in water and filter from insoluble material. The determination could be done by hplc, but there are titrimetric or electrochemical methods that would be easier.

(*iii*) The caffeine may have to be separated from flavours, colouring materials and other additives present in the drink. For this one, hplc would be a suitable technique.

(*iv*) and (*v*) Both of these could be separated by gc or by hplc, but hplc would probably be the better technique in both examples. The sugars could not be volatilised without decomposition, so by gc they would have to be examined as volatile TMS (trimethylsilyl) derivatives. The amines are polar substances that would show pronounced tailing by gc and would have to be derivatised as well. Both could be examined by hplc without pretreatment.

****

SAQ 2.3a

In Section 2.3.3 you worked out the effect of extra column dispersion on the peak of an unretained solute, using a column with a plate number of 10 000. The extra-column dispersion will decrease the plate number that we actually observe for this column. The table below contains the retention volume, peak widths and plate number for an unretained solute on this column.

$\longrightarrow$

SAQ 2.3a
(cont.)

(*a*) Complete the table by calculating the corresponding values for columns with plate numbers of 5000 and 3000 respectively.

(*b*) What is the % reduction in the plate number of each column due to extra-column effects?

N (ideally)	10 000	5000	3000
V_R, μl	2908.5	2908.5	2908.5
w_B, μl	116		
w_T, μl	126		
N (actual)	8525		
% reduction in N			

Response

N (ideally)	10 000	5000	3000
V_R, μl	2908.5	2908.5	2908.5
w_B, μl	116	165	212
w_T, μl	126	172	218
N (actual)	8525	4573	2847
% reduction in N	15	8.5	5

In each case the plate numbers are calculated using Eq. 2.3a and the actual peak width w_T. Thus, for the first column:

$$N = 16 \times (29085)^2/(126)^2 = 8525$$

You can see that the effect of a given amount of extra-column dispersion is more serious the higher the efficiency of the column.

****

SAQ 2.3b

Suppose you are running a 4.6 mm hplc column on a mixture of acetonitrile and water (80% by volume acetonitrile). The column runs continuously for 8 hours a day at a flow rate of 2 cm^3 min^{-1}, and your acetonitrile costs you seven pound per dm^3.

(a) What is the cost of acetonitrile per year, assuming a year = 250 working days?

(b) What is the mobile phase velocity (cm min^{-1}) through the column?

(c) If you changed to a 1 mm column operated at the same velocity, what flow rate would you have to use?

(d) What would the small bore column save you in acetonitrile costs?

Response

(a) The column uses $2 \times 60 \times 8 = 960$ cm^3 of mobile phase per day or $960 \times 0.8 = 768$ cm^3 of acetonitrile per day or 192 dm^3 per year, which costs one thousand, three hundred and forty four pounds.

(*b*) $2 = \pi \times (0.46)^2 \times v/4 \qquad v = 12.03$ cm min^{-1}.

(*c*) $f = \pi \times (0.1)^2 \times 12.03/4 \quad f = 0.0945$ cm^3 min^{-1}.

(*d*) acetonitrile cost $= 1344 \times 0.0945/2 = £63.5$, so the saving is £1280.5 per year.

****

SAQ 2.4a

Infrared absorption detectors are available for hplc, although they have never become very popular. From what you know about ir spectrometry and what you have read so far about hplc detectors, see if you can decide whether the following statements are true or false.

(*i*) An infrared spectrum provides more structural information about a compound than does a uv spectrum.

(*ii*) An ir detector would be more sensitive than a uv detector.

(*iii*) An ir detector could not be used with solvent mixtures containing water.

(*iv*) An ir detector could be used as a selective detector or a universal detector, by changing the wavelength used.

Response

(*i*)   T.

(*ii*)   F. If you have ever done any practical ir work you should have had no trouble with these two. An infrared spectrum provides a great deal of structural detail, but the method does not have very high sensitivity.

(*iii*) F. The flow cell would have to be made from a water insoluble material that is transparent to infrared radiation, eg KRS5 (TlBr/TlI). Glass cannot be used for optical components in ir instruments, as it absorbs ir radiation.

(*iv*) T with qualifications. The detector could be used selectively by operating at, for instance, 1725 cm^{-1}, when it would detect some solutes containing carbonyl groups, or as a universal detector at somewhere in the CH stretching region. The problem with universal detection would be to find a wavelength that was not absorbed by the mobile phase. The detector would have to be operated in 'windows' in the ir spectrum of the mobile phase; the number of suitable solvents is very limited.

****

SAQ 2.4b

Fig. 2.4f shows the uv spectra of azobenzene (Az, concentration $3.73 \times 10^{-3} \text{ g dm}^{-3}$) and phenanthrene (P, $3.23 \times 10^{-3} \text{ g dm}^{-3}$) both recorded in *iso*-octane. The wavelength drive on the instrument was 10 nm cm^{-1} and the absorbance range was 2 aufs. Measurements were made against *iso*-octane using 10 mm cells.

What wavelength would you choose:

(*i*) To detect Az without detecting P

(*ii*) To detect P without detecting Az

(*iii*) To detect both of them

(*iv*) To detect Az at maximum sensitivity?

$\longrightarrow$

SAQ 2.4b (cont.)	Calculate the molar absorptivity of each of them (in $dm^3\ mol^{-1}\ cm^{-1}$) at the wavelength you chose in (iv).

Response

(i)   The shoulder at 342 nm

(ii) It cannot be done, but at 251 nm the ratio of P to Az sensitivity would be greatest

(iii) 300 nm

(iv) Az has a maximum absorbance at 314 nm; at this wavelength:

	P	Az
Absorbance	0.17	0.55
M_r	178	182
c (mol dm^{-3})	1.81×10^{-5}	2.05×10^{-5}
$\epsilon = A/c$, $dm^3\ mol^{-1}\ cm^{-1}$	9.39×10^3	2.68×10^4

****

SAQ 2.4c	Fig. 2.4l shows the current-potential curves of two electroactive solutes X and Y. In a solution containing both of them: (i) Which would be detected by an ec detector operating at a potential E_2. $\longrightarrow$

SAQ 2.4c
(cont.)

> (*ii*) Which would be detected at E_3?
>
> (*iii*) Operating the detector at a potential E_4 would not be a good idea, and operation at E_1 would not be very smart, either. What would be detected at each of these potentials, and what is wrong with the choice of potential in each case?

Response

Operation at E_2 would detect X, operation at E_3 would detect both X and Y. Operation at E_4 would detect both X and Y, but at this potential the solvent or background electrolyte is oxidised as well. At best, there would be a large background current; it might be impossible to get the recorder on scale. At E_1, X would be detected, but the sensitivity would be low. It would be much better to work at a potential on the limiting current plateau, such as E_2.

****

SAQ 2.4d

> For which of the following analyses do you think that uv absorbance detection would not be suitable? If uv absorbance is unsuitable, suggest an alternative detector.
>
> (*i*) The determination of mixed sulphonamide drugs in a tablet.
>
> (*ii*) The separation of poly-ethene into fractions of different relative molecular mass, using exclusion chromatography. $\longrightarrow$

**SAQ 2.4d
(cont.)**

(*iii*) The determination of phenols as contaminants in a sample of river water.

(*iv*) The analysis of B-vitamins in a multivitamin tablet.

(*v*) The determination of riboflavin (vitamin B2) in milk.

The general structure of the sulphonamides is:

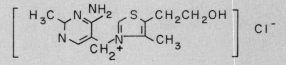

Structures of some B vitamins:

Thiamine(B1)

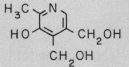

Pyridoxine(B6)

Niaciniamide

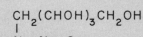

Riboflavin(B2)

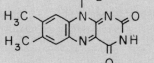

Response

(*i*)   These are aromatic, and so will absorb uv radiation. In a tablet, there should be reasonably high levels of the compounds present, so uv absorbance detection would be the method of choice.

(*ii*)   These compounds are saturated and will not show any uv absorption above 200 nm. A refractometer is the only suitable detector.

(*iii*)   The amounts of phenols present are likely to be very low. Trace phenols in water have been determined using both uv absorbance and ec detectors. The sensitivity of the uv absorbance detector is not really high enough, so that sample preconcentration methods have to be used. With the more sensitive ec detection, the analysis can be done without preconcentration.

(*iv*)   The structure of the vitamins indicates that they will all absorb uv radiation, so that for reasonably high levels of them in a tablet, uv absorbance detection would be suitable.

(*v*)   For riboflavin in milk, the low level of vitamin present might be a problem using a uv absorbance detector. Riboflavin has a highly conjugated structure, and will fluoresce, so that fluorescence detection could be used for trace amounts of this compound. The ring nitrogens in the structure indicate that ec detection would be possible as well.

****

SAQ 2.4e   Consider a solute which is detected by derivatisation, using a post-column reactor of the type
shown in Fig. 2.4p(i).

What would be the effect on the peak area of
this solute of:

(i)   Increasing the length of the reactor coil?

(ii)   Increasing the temperature of the reactor
coil?

(iii) Increasing the flow rate of reagent into the
mixing tee?

What would be the effect on the resolution between two peaks in the chromatogram of increasing the length of the reactor coil?

Response

(i)   Unless the derivatisation reaction is very fast, an increase in
the length of the reactor coil should increase the peak area,
because the longer the time the solute spends in the reactor,
the more product should be formed.

(ii) An increase in the temperature of the reactor should increase
the reaction rate and thus increase the peak area.

(iii) Changing the flow rate of the reagent may affect the peak area
for several reasons. At very low flow rates there may be in-
sufficient reagent to complete the reaction, in which case an
increase in flow rate will increase the peak area. But as we
increase the flow rate we are diluting our solute and also reducing the time the solute spends in the reactor, and these
factors may reduce the response at high flow rates.

An increase in the length of the reactor tube will increase the dispersion and so decrease the resolution between a given pair of solutes. In practice, the length of the reactor coil used will represent a compromise between detector response and resolution.

****

SAQ 3.2a

A test mixture consisting of phenyl methyl ketone, nitrobenzene, benzene and methylbenzene is to be separated on a C-18 column with a mobile phase of CH_3OH/H_2O 60:40. With these conditions, the ketone is eluted first.

(*i*)   In what order are the other solutes eluted?

(*ii*) How would you change the composition of the mobile phase so as to increase the retention of the solutes?

(*iii*) How would the retention of the solutes be affected by using a phenyl bonded phase instead of the C-18?

(*iv*) If the C-18 bonded phase contained unreacted silanol groups, how would the retention of the solutes be affected by end-capping the stationary phase?

Response

(*i*)   The more polar the solute, the faster it is eluted, so the order would be phenyl methyl ketone, nitrobenzene, benzene, methylbenzene.

(*ii*) We need to increase the polarity of the mobile phase, so we should increase the amount of water, provided that the mixture will still dissolve.

(*iii*) The phenyl bonded phase is slightly more polar than the C-18, and this will cause the solutes to elute faster, though it may not affect all the solutes to the same extent.

(*iv*) End-capping will slightly increase the percentage of carbon in the stationary phase, and will make it less polar. We would expect this to increase retention, and this is what happens for the non-polar solutes benzene and methylbenzene. For the other two, there is another effect to consider, because by end-capping we are reducing the adsorption of the polar solutes onto the stationary phase, and this will decrease their retention. In fact, for the two polar solutes there is a small decrease in retention on end-capping.

****

SAQ 3.3a   Explain the method used in each of the following:

(*i*) Separation of aspirin and norephedrine (1-phenyl,2-aminopropanol)

Column: C-18; mobile phase: CH_3OH/H_2O 50:50 + heptane sulphonic acid (pH about 3.5).

(*ii*) Chromatography of 4-aminobenzoic acid

Column: C-18; mobile phase: CH_3OH/H_2O 50:50 + tetrabutylammonium hydroxide (pH about 7.5).

Response

(*i*) Ion-pairing/ion-suppression. The aspirin is unionised in the acid solution, whereas the norephedrine, which is a weak base, will be fully protonated and is chromatographed as a neutral ion-pair.

(*ii*) Ion-pairing. The amine functionality is not protonated at pH 7.5 whereas the carboxylic acid function is fully ionised and pairs with the heptane sulphonic acid.

****

SAQ 3.4a

The following statements refer to the different modes of hplc. Indicate whether the statements are true (T) or false (F).

(*i*) In adsorption chromatography, a non-polar mobile phase is used.

(*ii*) Polar molecules can easily be separated by adsorption chromatography.

(*iii*) The retention times of solutes in exclusion chromatography can be altered by changing the polarity of the mobile phase.

(*iv*) Exclusion chromatography is useful only for the separation of large molecules.

(*v*) In reverse phase chromatography, the mobile phase is more polar than the stationary phase.

(*vi*) In reverse phase chromatography using bonded silica packings, the bonded group is non-polar.

Response

(*i*)   T. The mobile phase should be relatively non-polar

(*ii*) F. Polar molecules are best separated by reverse phase techniques. Using adsorption, polar molecules have long retention times and suffer from tailing.

(*iii*) F. Provided that exclusion is the only effect that is operating and provided that a change in the mobile phase does not affect the shape of the solute molecule, a change in polarity of the mobile phase will have little effect, as retention in this case is determined by the size and shape of the molecule in relation to the pore size of the stationary phase.

(*iv*) F. Exclusion chromatography is often used to separate small molecules from larger molecules.

(*v*)   T.

(*vi*) F. The bonded group is usually non-polar, but the condition for reverse phase operation is given by statement (*v*). Polar bonded groups can be used, provided the mobile phase is more polar than the stationary phase.

****

SAQ 3.4b	Which mode of hplc would you choose for each of the following: (*i*) Identification of plasticisers in polychloroethene (polyvinyl chloride). Common pvc plasticisers are dibutyl, dioctyl and dinonyl phthalates. (*ii*) Separation of tranquillisers. $\longrightarrow$

SAQ 3.4b (cont.)

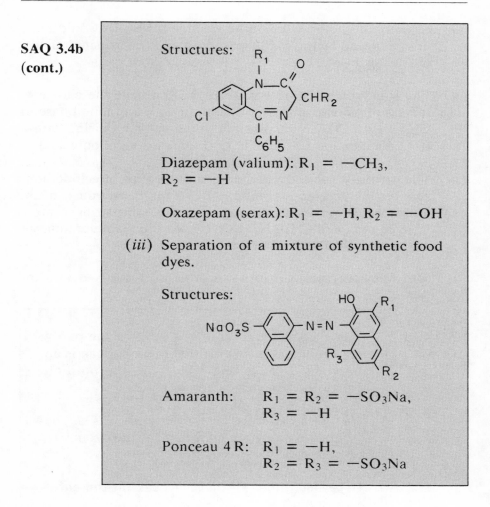

Structures:

Diazepam (valium): $R_1 = -CH_3$, $R_2 = -H$

Oxazepam (serax): $R_1 = -H$, $R_2 = -OH$

(*iii*) Separation of a mixture of synthetic food dyes.

Structures:

Amaranth: $R_1 = R_2 = -SO_3Na$, $R_3 = -H$

Ponceau 4 R: $R_1 = -H$, $R_2 = R_3 = -SO_3Na$

Response

(*i*) Phthalates can be separated by either normal or reverse phase techniques (see for example Fig. 4.3k), but if you wanted to do this you would either have to extract the plasticisers from the pvc or else risk contaminating your column with high relative molecular mass material. With the correct choice of exclusion column, the polymer could be excluded, when it would elute first, followed by the phthalates in order of decreasing relative molecular mass.

(*ii*) The tranquillisers are weak bases, so you have the choice of ion-suppression using a C-18 column with a mobile phase of CH_3OH or CH_3CN + alkaline buffer; or ion-pairing with a C-18 column and CH_3OH/H_2O + an alkyl sulphonic acid.

(*iii*) The strongly ionised sulphonic acid groups preclude ion-suppression for these compounds, so for these you can use ion-exchange on an anion-exchanger, or ion-pairing using a C-18 column and CH_3OH/H_2O + a tetrabutylammonium salt as the mobile phase.

****

SAQ 4.1a

The chromatogram in Fig. 4.1c has some partly resolved peaks. Assume that the first peak is an unretained solute and take V_o as the position where the first peak starts to elute.

(*i*) Draw a scale on the abscissa axis of the chromatogram marking k' values from 0 to 5.

(*ii*) Determine k' for each peak on the chromatogram.

(*iii*) Measure peak widths and calculate the resolution (R_S) between peaks 1 and 2, peaks 2 and 3 and peaks 4 and 5. For the peaks that are partly resolved you will have to extrapolate the linear part of the side of each peak down to the baseline, as in Fig. 4.1a.

$\longrightarrow$

SAQ 4.1a
(cont.)

> (*iv*) Calculate the selectivity (α) for peaks 2 and 3, 3 and 4, 4 and 5.
>
> (*v*) For the last two peaks, find the plate number and plate height of the column using Eq. 2.3c (the column is 25 cm long and has a relatively low efficiency).
>
> (*vi*) Tabulate the results.

Response

(*i*)

k'	0	1	2	3	4	5
distance from injection	12	24	36	48	60	72 mm

		Peak No			
	1	2	3	4	5
V_R, mm	14.5	18	24	43	70
(*ii*) k'	0.2	0.5	1	2.6	4.8
w_B, mm	3	3.5	3.5	4.5	7
$w_{\frac{1}{2}}$, mm				2.5	3.5
(*iii*) R_S		1.1	1.7		4.7
(*iv*) α			2	2.6	1.8
(*v*) N				1638	2216
H, mm				0.15	0.11

****

SAQ 4.3a   Suggest a gradient that would improve each of the chromatograms in Fig. 4.3k. You do not need to worry about the detail, like the exact shape of the gradient, or how long it will take. Concentrate on the composition of mobile phase that is needed at the start and at the end of each chromatogram.

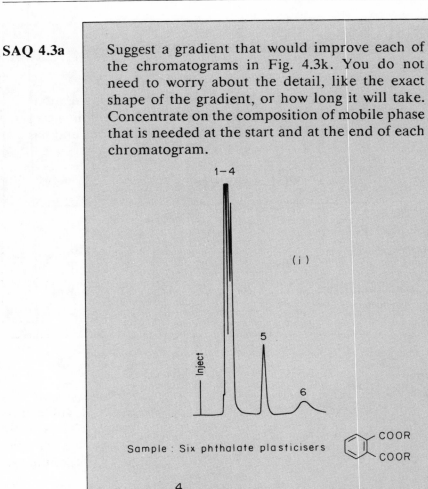

Sample : Six phthalate plasticisers

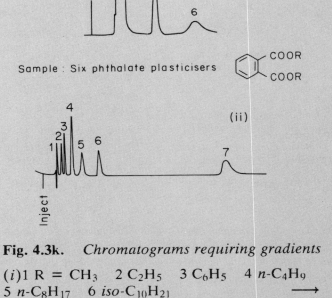

Fig. 4.3k.   *Chromatograms requiring gradients*

(i)1 R $=$ CH$_3$ 2 C$_2$H$_5$ 3 C$_6$H$_5$ 4 n-C$_4$H$_9$
5 n-C$_8$H$_{17}$ 6 iso-C$_{10}$H$_{21}$ $\longrightarrow$

SAQ 4.3a (cont.)

Column:	10 μm C-18 bonded phase, 30 cm × 4 mm
Mobile phase:	methanol/water 90 : 10. Flow rate 2 cm^3 min^{-1}
Detector:	Uv absorption, 254 nm

(*ii*)

Sample:	1 benzene 2 diphenyl ether 3 ethyl benzoate 4 carbazole 5 nitrobenzene 6 diphenyl ketone 7 benzyl alcohol
Column:	5 μm C-18 bonded phase. 30 cm × 1.8 mm
Mobile phase:	dichloromethane/hexane 40 : 60. Flow rate 2 cm^3 min^{-1}
Detector:	Uv absorption, 254 nm.

Response

(*i*) For the reverse phase separation (chromatogram (*i*)) the mobile phase needs to be more polar at the start and less polar at the end, so you should start with a mobile phase containing more water, say, methanol/water 75 : 25, and programme to pure methanol.

(*ii*) For the adsorption column (chromatogram (*ii*)) we want the mobile phase to be less polar at the start and more polar at the end, so we would start with a proportion of dichloromethane in hexane lower than 40% and finish with a concentration higher than 40%.

To sort out the best conditions, you would need to do some experimental work. Fig. 4.3l (*i*) shows the reverse phase separation run with the suggested gradient. In Fig. 4.3l (*ii*) the normal phase separation starts with 10% dichloromethane in hexane. This is run isocratically for 3 minutes, then the proportion of dichloromethane is increased at 8% min to 40%. At this point the programming rate was halved to 4% min, and the programme was continued to 100% dichloromethane.

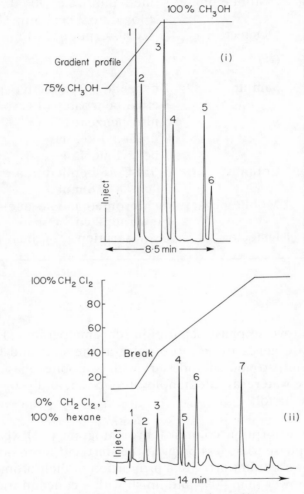

Fig. 4.31.   *Examples of gradients*

(*i*) Reverse phase separation of phthalates using a gradient.

(*ii*) Normal phase separation using a gradient (sample as in Fig. 4.3k (*ii*))

****

SAQ 4.4a

A commercial analgesic tablet is stated on the packet to contain 325 mg of aspirin and 50 mg of caffeine per tablet. Two tablets + 0.0773 g of phenacetin were shaken with 10 cm^3 of ethanol for 10 minutes, then 10 cm^3 of 0.5 mol dm^{-3} ammonium formate were added and the mixture made up to 100 cm^3. The tablets contain insoluble excipients, so a little of the solution was filtered before chromatography

Fig. 4.4f gives peak areas obtained for two 1 μl injections of the solution. Chromatographic conditions were as described in Section 4.4

Injection no.	Peak area		
	Aspirin	Phenacetin	Caffeine
1	157595	170804	50693
2	153541	164174	48478

Fig. 4.4f

Using the relative response factors that you calculated in the last Section, correct each peak area by dividing by the appropriate response factor. For each injection, calculate the amount of aspirin and caffeine present, expressing the results as mg per tablet.

The aspirin content should be between 95 and 105% and the caffeine content between 90 and 110% of the amount stated on the packet. Are the tablets within specification?

Response

No. 1	Aspirin	Phenacetin	Caffeine
peak area	157595	170804	50693
r	0.111	1	0.217
corrected area	1419775	170804	223608
normalised area, %	77.83	9.36	12.81
mass in mixture, mg	642.8	77.3	105.7
mg per tablet	321.3		52.9
No. 2			
Peak area	153541	164174	48478
corrected area	1383252	164174	223401
normalised area, %	78.11	9.27	12.62
mass in mixture, mg	651.3	77.3	105.2
mg per tablet	325.7		52.6

The allowed limits are: aspirin 309–341 mg per tablet

caffeine 45–55 mg per tablet

Hence the tablets are within specification.

****

SAQ 5.2a The following account, taken from a practical notebook, describes the preparation of the mobile phase used in the example in Section 4.4.

'1.575 g of ammonium formate was made up to 500 cm^3 in a graduated flask. To this solution, 50 cm^3 of ethanol was added, and after mixing the mobile phase was placed in the solvent reservoir and pumping was commenced at 2 cm^3 min^{-1}'.

Can you identify three mistakes that were made?

Response

(*i*) After adding the ethanol, the solution is no longer 0.05 mol dm^{-3} in ammonium formate, nor does it contain 10% of ethanol.

(*ii*) The mobile phase was not filtered before use.

(*iii*) The mobile phase was not degassed before use.

Another common error in the preparation of this mobile phase is the use of industrial grades of ethanol, which contain uv-absorbing impurities. The source of the ethanol should have been specified as well.

****

SAQ 5.2b

These questions refer to the properties of solvents that were listed in Fig. 5.2a.

(*i*)   Considering a heptane/trichloromethane mobile phase, why would it be undesirable to replace the heptane by pentane, or the trichloromethane by tetrachloromethane?

(*ii*)   Why are methanol or acetonitrile preferred to other water miscible solvents for the preparation of mobile phases for reverse phase chromatography?

(*iii*) What would be the major difficulty associated with the use of propanone in mobile phases?

Response

(*i*)   Pentane is more expensive than heptane and is also much more volatile. Control of the mobile phase composition might be difficult.

Tetrachloromethane has a higher viscosity than trichloromethane (leading to a greater pressure drop across the column). It is also toxic.

(*ii*) Methanol is relatively cheap, and has a low viscosity and low uv cut-off. Although more expensive and more toxic than methanol, acetonitrile has a lower uv cut-off. The others have higher viscosities, higher cut-offs and are more expensive than methanol. With ethanol, there are additional problems with Customs and Excise.

(*iii*) The major problem with propanone is the high uv cut-off.

****

SAQ 5.3a	These questions refer to the extraction and separation of azines, that was discussed in the previous Section:

(*i*) What difference between the azines and the PAHs is exploited to achieve their separation on the C-18 cartridge?

(*ii*) Why is HCl used when the cartridge is eluted?

(*iii*) If the PAHs had not been removed from the sample, how would you expect their retention times to compare with those of the corresponding azines (eg anthracene and acridine or naphthalene and quinoline)?

(*iv*) The chromatographic separation was done at pH 7.2. Can you think of any disadvantages of using a lower or a higher pH?

(*v*) Which detector could be used for the separation?

Response

(*i*) The azines are weakly basic compounds.

(*ii*) The weak bases are protonated by the HCl. The protonated bases are highly polar species which elute rapidly from the cartridge, whilst the non-polar PAHs are retained.

(*iii*) The PAHs would be strongly retained by the nonpolar stationary phase and their retention times would be much longer than those of the corresponding azines.

(*iv*) In acid buffers, the azines may be protonated. This would cause loss of efficiency and poor peak shape. The effect will diminish as the pH is increased, but at pH more than about 8, the lifetime of the column may be reduced, due to dissolution of the silica stationary phase.

(*v*) As with PAH, the conjugated structure of the azines suggests that fluorescence detection could be used. In fact, both fluorescence and uv absorption detection were used in this work.

****

SAQ 5.3b | Comparing on-line and off-line methods of column switching, which of the following advantages and limitations do you think would apply to the on-line technique?

(*i*) Easier to carry out

(*ii*) Cheaper

(*iii*) Faster

(*iv*) Better reproducibility

(*v*) More chance of sample loss

(*vi*) Easier to automate.

Response

On-line techniques are easily automated, but are more expensive as they require additional valves with associated switching equipment. Off-line methods are rather easier to carry out, but, because of the sample collection and re-injection steps, they are slower and tend to be less reproducible. With off-line techniques there is also more chance of sample loss due to adsorption or evaporation.

****

Units of Measurement

For historic reasons a number of different units of measurement have evolved to express quantity of the same thing. In the 1960s, many international scientific bodies recommended the standardisation of names and symbols and the adoption universally of a coherent set of units—the SI units (Système Internationale d'Unités)—based on the definition of five basic units: metre (m); kilogram (kg); second (s); ampere (A); mole (mol); and candela (cd).

The earlier literature references and some of the older text books, naturally use the older units. Even now many practicing scientists have not adopted the SI unit as their working unit. It is therefore necessary to know of the older units and be able to interconvert with SI units.

In this series of texts SI units are used as standard practice. However in areas of activity where their use has not become general practice, eg biologically based laboratories, the earlier defined units are used. This is explained in the study guide to each unit.

Table 1 shows some symbols and abbreviations commonly used in analytical chemistry; Table 2 shows some of the alternative methods for expressing the values of physical quantities and the relationship to the value in SI units.

More details and definition of other units may be found in the *Manual of Symbols and Terminology for Physicochemical Quantities and Units*, Whiffen, 1979, Pergamon Press.

Table 1 *Symbols and Abbreviations Commonly used in Analytical Chemistry*

Å	Angstrom
$A_r(X)$	relative atomic mass of X
A	ampere
E or U	energy
G	Gibbs free energy (function)
H	enthalpy
J	joule
K	kelvin (273.15 + t °C)
K	equilibrium constant (with subscripts p, c, therm etc.)
K_a, K_b	acid and base ionisation constants
$M_r(X)$	relative molecular mass of X
N	newton (SI unit of force)
P	total pressure
s	standard deviation
T	temperature/K
V	volume
V	volt ($J A^{-1} s^{-1}$)
$a, a(A)$	activity, activity of A
c	concentration/ mol dm^{-3}
e	electron
g	gramme
i	current
s	second
t	temperature / °C
bp	boiling point
fp	freezing point
mp	melting point
$\approx$	approximately equal to
$<$	less than
$>$	greater than
e, $\exp(x)$	exponential of x
$\ln x$	natural logarithm of x; $\ln x = 2.303 \log x$
$\log x$	common logarithm of x to base 10

Table 2 *Alternative Methods of Expressing Various Physical Quantities*

1. **Mass (SI unit : kg)**

$$g = 10^{-3} \text{ kg}$$
$$mg = 10^{-3} \text{ g} = 10^{-6} \text{ kg}$$
$$\mu g = 10^{-6} \text{ g} = 10^{-9} \text{ kg}$$

2. **Length (SI unit : m)**

$$cm = 10^{-2} \text{ m}$$
$$\text{Å} = 10^{-10} \text{ m}$$
$$nm = 10^{-9} \text{ m} = 10\text{Å}$$
$$pm = 10^{-12} \text{ m} = 10^{-2} \text{ Å}$$

3. **Volume (SI unit : m^3)**

$$l = dm^3 = 10^{-3} \text{ m}^3$$
$$ml = cm^3 = 10^{-6} \text{ m}^3$$
$$\mu l = 10^{-3} \text{ cm}^3$$

4. **Concentration (SI units : mol m^{-3})**

$$M = \text{mol } l^{-1} = \text{mol dm}^{-3} = 10^3 \text{ mol m}^{-3}$$
$$\text{mg } l^{-1} = \mu g \text{ cm}^{-3} = \text{ppm} = 10^{-3} \text{ g dm}^{-3}$$
$$\mu g \text{ g}^{-1} = \text{ppm} = 10^{-6} \text{ g g}^{-1}$$
$$\text{ng cm}^{-3} = 10^{-6} \text{ g dm}^{-3}$$
$$\text{ng dm}^{-3} = \text{pg cm}^{-3}$$
$$\text{pg g}^{-1} = \text{ppb} = 10^{-12} \text{ g g}^{-1}$$
$$\text{mg}\% = 10^{-2} \text{ g dm}^{-3}$$
$$\mu g\% = 10^{-5} \text{ g dm}^{-3}$$

5. **Pressure (SI unit : N m^{-2} = kg m^{-1} s^{-2})**

$$Pa = Nm^{-2}$$
$$\text{atmos} = 101\ 325 \text{ N m}^{-2}$$
$$\text{bar} = 10^5 \text{ N m}^{-2}$$
$$\text{torr} = \text{mmHg} = 133.322 \text{ N m}^{-2}$$

6. **Energy (SI unit : J = kg m^2 s^{-2})**

$$\text{cal} = 4.184 \text{ J}$$
$$\text{erg} = 10^{-7} \text{ J}$$
$$\text{eV} = 1.602 \times 10^{-19} \text{ J}$$

Table 3 *Prefixes for SI Units*

Fraction	Prefix	Symbol
10^{-1}	deci	d
10^{-2}	centi	c
10^{-3}	milli	m
10^{-6}	micro	μ
10^{-9}	nano	n
10^{-12}	pico	p
10^{-15}	femto	f
10^{-18}	atto	a

Multiple	Prefix	Symbol
10	deka	da
10^{2}	hecto	h
10^{3}	kilo	k
10^{6}	mega	M
10^{9}	giga	G
10^{12}	tera	T
10^{15}	peta	P
10^{18}	exa	E

Table 4   *Recommended Values of Physical Constants*

Physical constant	Symbol	Value
acceleration due to gravity	g	9.81 m s^{-2}
Avogadro constant	N_A	$6.022\ 05 \times 10^{23} \text{ mol}^{-1}$
Boltzmann constant	k	$1.380\ 66 \times 10^{-23} \text{ J K}^{-1}$
charge to mass ratio	e/m	$1.758\ 796 \times 10^{11} \text{ C kg}^{-1}$
electronic charge	e	$1.602\ 19 \times 10^{-19} \text{ C}$
Faraday constant	F	$9.648\ 46 \times 10^{4} \text{ C mol}^{-1}$
gas constant	R	$8.314 \text{ J K}^{-1} \text{ mol}^{-1}$
'ice-point' temperature	T_{ice}	$273.150 \text{ K exactly}$
molar volume of ideal gas (stp)	V_m	$2.241\ 38 \times 10^{-2} \text{ m}^3 \text{ mol}^{-1}$
permittivity of a vacuum	ϵ_0	$8.854\ 188 \times 10^{-12} \text{ kg}^{-1} \text{ m}^{-3} \text{ s}^4 \text{ A}^2 \text{ (F m}^{-1})$
Planck constant	h	$6.626\ 2 \times 10^{-34} \text{ J s}$
standard atmosphere pressure	p	$101\ 325 \text{ N m}^{-2} \text{ exactly}$
atomic mass unit	m_u	$1.660\ 566 \times 10^{-27} \text{ kg}$
speed of light in a vacuum	c	$2.997\ 925 \times 10^{8} \text{ m s}^{-1}$